Exposing Electronics

Artefacts: Studies in the History of Science and Technology

In growing numbers, historians are using technological artefacts in the study and interpretation of the recent past. Their work is still largely pioneering, as they investigate approaches and modes of presentation. But the consequences are already richly rewarding. To encourage this enterprise, three of the world's great repositories of the material heritage of science and technology: the Deutsches Museum, the Science Museum and the Smithsonian Institution, are collaborating on this new book series. Each volume will treat a particular subject area, using objects to explore a wide range of issues related to science, technology and medicine and their place in society.

Edited by Robert Bud, Science Museum, London
Bernard Finn, Smithsonian Institution, Washington
Helmuth Trischler, Deutsches Museum, Munich

Volume 1 Manifesting Medicine: Bodies and Machines

Principal Editor Robert Bud

Volume 2 Exposing Electronics

Principal Editor Bernard Finn

Further volumes in preparation, on the themes of:

Transport
Communication
Environment

This book is part of a series. The publisher will accept continuation orders which may be cancelled at any time and which provide for automatic billing and shipping of each title in the series upon publication. Please write for details.

Exposing Electronics

Edited by

Bernard Finn
Smithsonian Institution, Washington, DC

Associate Editors

Robert Bud
Science Museum, London

Helmuth Trischler
Duetsches Museum, Munich

harwood academic publishers
Australia • Canada • France • Germany • India
Japan • Luxembourg • Malaysia • The Netherlands
Russia • Singapore • Switzerland

 Printed in Singapore.

Amsteldijk 166
1st Floor
1079 LH Amsterdam
The Netherlands

British Library Cataloguing in Publication Data
A catalogue record for this book is available from the British Library.

ISBN: 90-5823-056-2 (hard cover)
ISBN: 1029-3353

Contents

Bell-Gray comparison. Photo by Laurie Minor

Illustrations

Series Preface

In the long history of the efforts made by science museums to promote the importance of their collections, the past decade has been among the most exciting. Whereas the competition from non-object based science centres has become ever stronger, interest in using objects to communicate insight into the history of our technological and scientific heritage has gained new strength. For millions of visitors, artefacts provide a uniquely attractive and direct link to the past.

Museums also have a research mission. They are a vital force in the community of scholars, especially in the history of technology, and here, too, they have come to be better appreciated. Many outside their walls have come to share the belief that artefacts have played a role which is both inadequately understood and indispensable for a better understanding of historical and cultural change.

Initially, perhaps, it was the insight into technical detail provided by close inspection of the real thing that was generally of greatest scholarly importance. More recently, however, studies of experiments and technology have widened the view to the complex role of artefacts within their larger geographical, economic, social and political setting. Rather than being treated in isolation, technological objects and instruments are coming to be used as material expressions of human culture that shape, mediate and reflect the interactions amongst science, technology and society. Latter-day onlookers are therefore helped to see not just machines, but also imaginative worlds of the past.

Building on rapidly maturing scholarly interest, three of the world's great repositories of material heritage (the Deutsches Museum in Munich, the National Museum of American History in Washington and the Science Museum in London) are cooperating to support this new series of publications. Volumes will explore innovative approaches to the object-oriented historiography of science and technology. The series will seek to go beyond a strict technical description of artefacts on the one hand, and an overly broad social history on the other.

Collections reflect local, regional and national traditions and express their cultures and history. This character confers certain constraints, but also advantages. Museums are sensitive to, and reflect, the specific local meanings of objects, but they have the asset, too, of curators whose detailed knowledge of the collections is couched within a wider historical perspective.

Building on these dual strengths, the series is intended to initiate an international discussion which both emphasizes local material cultures, and also draws upon recent research in the overall history of science and technology. The authors will therefore include curators, but the series will attract into the discussion other scholars from a much wider orbit. Many people have, of course, been concerned with the problems examined in this series; but all too often this has been in individual or institutional isolation. These volumes will engage an international community that is large enough to develop research programmes and debates that will have enduring momentum and excitement.

Situated at the interface between museum, university and independent research institution, the series will address professional historians of science and technology, curators, those in charge of the day-to-day administration of museums and those who, so often passionately, simply enjoy visiting. As museums do in general, the series aims to build a bridge between historical research and the use and application of historical knowledge in education and the public understanding of science and technology.

Each volume will focus on a specific field of technology and science in its wider historical context. The first, and larger, part of each volume will present the honed products of presentation and debate at joint conferences. The second part will consist of exhibit reviews, critical expeditions into the respective museum's landscape, bibliographical overviews on recent literature, and the like.

The collaboration between three national institutions has been made possible by their directors. We thank Neil Cossons, Director of the Science Museum, Spencer Crew, Director of the National Museum of American History of the Smithsonian Institution, and Wolf Peter Fehlhammer, Director of the Deutsches Museum. Their personal enthusiasm for this project has made it possible.

This series has also depended on the staff of Harwood Academic Publishers for their engaged and passionate interest.

Notes on Contributors

Ross Bassett is an assistant professor of history at North Carolina State University. His dissertation on the history of the MOS transistor won the 1998 Krooss Prize from the Business History Conference.

Roger Bridgman is Curator of Communications at the Science Museum, London, but currently working full time on exhibition content for the digital technology gallery of the museum's new Wellcome Wing. Research interests include the use of instruction leaflets as a source for the history of technology.

Paul Ceruzzi is Curator of Aerospace Electronics and Computing at the Smithsonian's National Air and Space Museum. His most recent book is *A History of Modern Computing*, and his latest exhibition deals with the Global Positioning System; both appeared in 1998.

Jon Eklund is Curator Emeritus at the Smithsonian's National Museum of American History. His former responsibilities were for chemistry and computers and was a co-curator of the Information Age exhibit. He is presently completing documentation of the chemistry collections.

Bernard Finn is Curator of the electrical collections at the Smithsonian's National Museum of American History, USA. Current research interests include electric lighting, submarine telegraphy, and the history of technical museums. He was a co-curator of the Information Age exhibit.

Paul Forman has been curator for modern physics at the Smithsonian Institution's National Museum of American History, since 1973. His research has dealt largely with the cultural and institutional context of the pursuit of physics, and with the influence of such context upon the constructs of that science.

Sungook Hong teaches the history of physics and engineering at the Institute for the History and Philosophy of Science and Technology, University of Toronto. He is interested in the history of wireless telegraphy, power engineering, and the relationship between science and technology in their developments.

Kirk Jeffrey is a Professor of History at Carleton College, Minnesota. His book on the invention of the cardiac pacemaker and the implantable defibrillator, and the subsequent growth of the rhythm management industry will be published in 2001.

Alan Q. Morton is Acting Head, Physical Sciences and Engineering Group, at the Science Museum, London. Currently he is leading a project to develop a new exhibition on energy. His research interests are in the area of energy policy.

Harmut Petzold is curator for computer science and also for time measurement at the Deutsches Museum. His research interests involve the history of information technology.

David Rhees is Executive Director of the Bakken Library and Museum in Minneapolis. His current research concerns the development of therapeutic medical technologies, for which he is conducting oral histories with some of the pioneers of the medical device industry in Minnesota.

Bernard Finn

Introduction

In the first volume of *Artefacts: Manifesting Medicine,* we demonstrated some of the ways that object-driven research can provide fresh insights into the history of medicine and health. Here, in the second volume, we turn to electronics, with a rich diversity of approaches—about which more shortly.

But first I shall make some more general comments about the task that we have undertaken. Our goal is to persuade other historians that artefacts are fruitful sources of inspiration and of evidence, which might help persuade them to pay more attention to the collections we have so carefully accumulated in our museums. We also want to provide examples of how to write about those artefacts in a historically fruitful manner.

It isn't easy. Archeologists provide a model of sorts. For them, objects are the main—sometimes the only—evidence to be considered, and a whole system of interpretive procedures has emerged which, in effect, defines the discipline. But as we move forward in time, into periods where artefacts compete with written records, there is an increasing tendency to rely on the word instead of the object. And that is the way historians are trained. They use books and business documents and letters, and in the process learn how to use libraries and archives and microfilm readers and even computers. They employ prescribed forms of footnotes to provide readers with explicit references, and thus the opportunity to confirm statements. There is no room here for the object.

Each of our museums has, in the past, sponsored publications related to its collections. At the Smithsonian Robert Multhauf pressured, cajoled and otherwise stimulated members of the staff to produce a number of articles and short monographs which appeared as *Contributions from the Museum of History and Technology*; and in 1966 Walter Cannon was the founding editor of the *Smithsonian Journal of History*, which for three years accepted manuscripts from a broad range of authors.

In *Artefacts* we build on these earlier traditions, but we are also more explicit in wanting to encourage the development of a historiography that includes objects. We have a format that allows for the use of illustrations (a partial equivalent to the footnotes employed for text references), and as editors we do whatever we can to stimulate potential authors to think in this direction. With that in mind, our museums sponsor an annual meeting where ideas and sometimes whole papers emerge. Three major articles in this volume were the product of such meetings (the one by Jeffrey and Rhees from a session at the Science Museum in 1996, those by Petzold and Morton from a meeting at the Smithsonian in 1997). I hasten to add that they, like the other contributions, were all subjected to normal peer review.

Let me now take the opportunity to share with you some suggestions for how artefacts can be significant to historians. Most obvious, perhaps, is the use of the object as inspiration. We look at a Hershel telescope or a Watt engine or a de Forest audion and think 'isn't this interesting, I wonder how it came to be?' and promptly turn to traditional documents for the answers. The use of the object in this instance is not trivial. It provides a visual (and perhaps tactile) stimulus that is absent in a written or pictorial description. It also provides an emotional link that even for the jaded historian is not insignificant. There are many examples of this approach, including Kim Pelis' article on blood transfusion in *Manifesting Medicine*—the author clearly having been fascinated by early 19th century blood-collecting apparatus in the Science Museum collections.

But the object may also have been inspiring to someone else, at some earlier point in time, thus making it and the reaction to it the stuff of history. Here again there are many examples. My thoughts are drawn to a conference inspired by the centennial of the Brooklyn Bridge (an artefact of majestic proportions) in 1984, which resulted in several articles, including, for example, Alan Trachtenberg, 'Brooklyn Bridge as a Cultural Trust,' in M. Latimer, B. Hindle and M. Kranzberg (eds.), *Bridge to the Future: A Centennial Celebration of the Brooklyn Bridge* (New York, 1984).

The historian may want to consider the materials from which the object is made—information often lacking in or at odds with the literary sources. Robert Gordon has used his metallurgical skills to produce several revealing works, among them 'Material Evidence of the Development of Metal-Working Technology at the Collins Axe Factory,' *Journal of the Society for Industrial Archaeology* 9 (1983), 19–28.

Operation of the artefact can sometimes lead to surprising results. When John White revived the Smithsonian's locomotive 'John Bull' for its 150th birthday he found the ride smoother and the mechanism much more forgiving than anticipated—as can be seen in his *The John Bull: 150 Years a Locomotive* (Washington, 1981). And when I worked with some of Bell's early telephones I discovered that electrochemical decomposition had probably been the reason—left unexplained in his notebooks—that he so quickly abandoned the variable-resistance transmitter. For this see 'Alexander Graham Bell's Experiments with the Variable Resistance Transmitter,' *Smithsonian Journal of History* 1 (Winter 1966–67), 1–16. Another form of operation can tell us about the capabilities of an instrument, and therefore about the limitations on its users. A classic study in this area is Gerard Turner's 'An Electron

Microscopical Examination of Nobert's Ten-Band Test-Plate,' *Journal of the Royal Microscopical Society* 84 (April 1965), 65–75.

The object may yield clues through evidence of the way it was constructed. Thus some of the myths surrounding the history of 'interchangeable parts' were exploded by Edwin Battison in 'Eli Whitney and the Milling Machine,' *SJH* 1 (Winter 1966), 9–34 and by David Hounshell in Appendix 2 ('Singer Sewing Machine Artifactual Analysis') to his *From the American System to Mass Production, 1800–1932* (Baltimore, 1984), pp. 337–344. More recently, Rolf Willach has caused us to revise our notions of how John Dollond developed his achromatic lens in 'New Light on the Invention of the Achromatic Telescope Objective,' *Notes and Records of the Royal Society of London*, 50 (1996), 195–210.

Sometimes evidence of use—wear and tear—can provide significant historical clues. Again, Robert Gordon, in his 'Laboratory Evidence of the Use of Metal Tools in Machu Picchu and Environs,' *Journal of Archaeological Science* 12 (1985), 311–327.

Where we find things, how many we find, the serial numbers—this is the kind of evidence that can tell us about the popularity of a product, how many were sold and who was buying them. Deborah Warner used this technique in her pursuit of a particularly elusive instrument in 'Davis Quadrants in America,' *Rittenhouse* 3 (1988), 23–40.

Design elements have been used to draw conclusions about the 'style' of invention. Reese Jenkins applied this line of reasoning in his 'Elements of Style: Continuities in Edison's Thinking, in *Bridge to the Future*, pp. 149–162, though, unfortunately for us, he was working from drawings instead of objects. Steven Lubar uses artefacts for this purpose in considering John Howe in his 'Cultural and Technological Design in the Nineteenth Century Pin Industry,' *Technology & Culture* 28 (1987), 253–82.

Finally there is the matter of physical appearance—especially size and weight—which can make a strong impression on the historian (as on anyone else). The influence may be too subtle to register in a citation, but it is nevertheless something we should be aware of.

Now to more urgent matters: volume II of *Artefacts*. The choice of subject was determined at the London meeting of 1996 and was explored in several presentations and discussions in Washington the following year. Electronics, like Medicine in our first volume, obviously embraces a very broad territory. We were anxious to provide a feeling for that breadth and, through a combination of design and happenstance, I believe have managed to do so. And objects are at the focus of each contribution, although not covering the breadth of possibilities discussed above.

Sungook Hong appropriately starts us with a perceptive essay on how, or we might better say why, the electronic age began. The objects to which he draws our attention are the early Fleming valves which are preserved at the Science Museum; their function here is largely one of inspiration. Hong's message is yes, Fleming was directly influenced by Edison; and yes, his 'valve' emerged as a practical electron-manipulating device that triggered all that followed. But his motivation for and understanding of what he was doing are seen as quite different from the traditional account. Call it social determinism if you like; I see it as a wonderful example of the human nature of inventors as they pursue a technological course.

Alan Morton looks at the scientific side of electronic origins, examining the inspiration that J. J. Thomson's cathode ray tube provided for an earlier generation of would-be historians. Here there is no questioning of what J. J. Thomson did, or why he did it, but rather an intriguing exposure of how the physics community (with help from others, including Thomson himself) manipulated the memory of his experiments to establish a positive image for its discipline. Providentially, museums figure prominently in the story.

Skipping the major early-twentieth century applications of these events, Hartmut Petzold demonstrates how strongly the need was felt in the 1930s for a machine that electronics would eventually make possible. He examines several computational devices in order to assure himself that he knows how they work. The frustration is almost palpable as we consider how Wilhelm Cauer and numerous contemporaries were struggling to solve complex mathematical problems with ingenious arrangements of wires and resistors and gears and punched paper. Some of them, of course, worked, and were put to practical use; but unfortunately this was not true for Cauer's. From all this activity one might think that in the immediate post-ENIAC era there would have been a much greater realization that the electronic solution would have far-reaching consequences.

Kirk Jeffrey and David Rhees take us back to the realm of symbolism, describing in detail how a small white device became a cultural icon for a corporation and for public and professional communities. In the course of this they also describe one of the earliest and most significant applications of the solid-state form of electronics, embodied in the transistor. The cardiac pacemaker was a remarkable achievement, all the more interesting because of the fortuitous encounter at a fortuitous moment between an electrical engineer and a heart surgeon, producing the kind of interaction that we find much more common—and even actively encourage—today.

Finally, we approach the end of a hundred years of electronics, and Ross Bassett examines that ultimate black box—the integrated circuit, and more specifically the microprocessor. It turns out that with a proper microscope the IC can be very revealing of its character. With the assistance of a view at that level, he delves into an exploration of motivations—both personal and corporate. He finds that what is virtually the same device my have different meanings—and hence different historical fates—to their inventors.

But there is more. We proceed now to a museum's-eye view of electrical technology. First, I and a colleague, Jon Eklund, describe what we thought we were doing when we used certain artefacts in a major Smithsonian exhibition on the Information Age, concluding that they at least had the potential of being quite effective in conveying portions of our story. Our Science Museum counterpart, Roger Bridgman, believes that it is difficult—indeed virtually impossible—for objects to convey any real meaning in a thematic exhibit except to other experts; and he concludes that we have done no better than anyone else in this regard. There seems to be a difference of opinion here, which the reader can ponder. Then please come to the museum (where the exhibit is likely to be available for a few more years) and make the determination for yourself.

Next, two other Smithsonian colleagues take a curator's view of specific artefacts. Paul Ceruzzi considers some computers designed by Seymour Cray and asks questions about style. Can there be a style to circuit design? Was this true of Cray? If so, is it observable in his machines? And what meaning does it have?

Paul Forman takes a longer look at a couple of seemingly insignificant items that came to us from the estate of I. I. Rabi. He establishes that they must have had symbolic meaning to Rabi himself and goes on to consider what that meaning was and how it is relevant to us as historians.

To close the volume I expound on an issue that has long fascinated me, the relationship between private and public collecting. This takes me into a specific consideration of electrical collections at the Smithsonian. I then move out again to a broad survey of electrical collections in the world's museums, which I hope will be of some value to people both inside and outside the museum community.

The exercise of editing this volume has helped me to achieve personally much of what we hoped to achieve collectively through *Artefacts*: a closer relationship with colleagues in other museums, and a better understanding of the value of the objects that we collect. I hope that you, the readers, may achieve something of the same feeling.

Sungook Hong

Inventing the History of an Invention: J. A. Fleming's Route to the Valve

Engineers often recall their inventions. These recollections provide important narratives for the history of technology. Historians, however, are well aware that these recollections should not be taken literally. Engineers' memories of what happened, say, twenty years ago are not always exactly correct. They sometimes put too much emphasis on the novelties they introduced, while devaluing others' contributions. Recollections may give the engineer a psychological satisfaction, or may provide an intellectual glue to hold the engineering community together, but they sometimes serve a more direct purpose. When the 'authorship' of a technology is at stake in a court, recalling how one invented, or made a contribution to the invention, the technology in dispute is an important strategy for patent litigation. In this case, the engineer may well stress radical differences of his technology from that of others. The engineer may also stress the origins of that technology in his previous work, providing a smooth continuity between his past work and the later invention. The simultaneous existence of discontinuity (from others' work) and continuity (to his own work) frequently characterizes an engineer's recollections on the invention of novel artifacts.[1]

We can find a similar plot in John Ambrose Fleming's recollection of the invention of the thermionic valve in 1904. In a series of well-known recollections, Fleming remembered that he transformed the Edison effect into the valve in 1904, when a sensitive signal detector was badly needed for wireless telegraphy. The Edison effect was a curious effect that Edison and his assistants discovered in the early 1880s inside a specially constructed light bulb. In 1889 Fleming performed a series of experiments on the Edison effect and conceptualized it in terms of unilateral conductivity in the vacuum inside the bulb. Some of the lamps that Fleming used in 1889 are shown in Figure 1 and Figure 2.

Unlike Edison who did not understand its underlying mechanism, Fleming investigated it and eventually transformed it into the valve, a sensitive and convenient signal detector for wireless telegraphy, in 1904. When he came up with the idea of utilizing unilateral conductivity for wireless detectors, he took one of the old lamps (Figure 1 and Figure 2), which he had used 15 years previously, from his cupboard and tested its performance as a wireless detector. By doing so, Fleming is said to have mediated between lamp and radio engineering, as well as between physics and technology. Yet Fleming began to recall his invention of the valve at a

Figure 1. A specially constructed lamp that Fleming used for his experiments on the Edison effect and unilateral conductivity in 1889. Picture source: Science Museum, London.

Figure 2. Other lamps that Fleming used in 1889. Note that the lamp in the middle has a zigzag wire anode, while the other lamps have a metal plate anode. He later recalled that it never occurred to him to put a metal plate and a zigzag wire in the same lamp and use the latter to control the electron flow from the filament to the metal plate as Lee de Forest did. The value that Fleming placed on these lamps as evidence for his claims to priority can be adduced from two letters from Fleming to Henry Lyons, Director of the Science Museum (preserved in the Archives of the museum), which were brought to my attention by Robert Bud. In one, dated 19 October 1925, he stated, 'You have such a valuable collection of original priceless apparatus that it ought to include the original *thermionic valves which are the* progenitors *of all others.' In the other, dated 25 October 1925, he wrote, 'I see no objection to this [exhibition] provided they are marked so as to make it clear that these valves are my original valves and the progenitors of all other thermionic valves of whatever kind.' Picture source: Science Museum, London.*

time when he and the Marconi Company were fighting over the patent on the vacuum tube with Lee de Forest who invented the audion (or the triode). To support the important claim that his valve, not de Forest's triode, was a breakthrough in the history of vacuum-tube technology, Fleming, I will assert, highlighted the inevitability and smoothness of the transition from the Edison effect to the valve, the importance of his scientific advisorship to Marconi as a context in which the invention was made, and a well-defined usage of the valve as a detector in wireless telegraphy in 1904.[2]

This paper aims critically to analyze Fleming's own narrative of his invention of the valve. Instead of stressing inevitability and continuity, I will focus on local and temporal complexities and contingencies specific to Fleming around 1904. I will also show that the termination of Fleming's scientific advisorship to the Marconi Company in December 1903, as well as his efforts to regain his connection to it, rather than his alleged scientific advisorship to Marconi, was a crucial factor that led him to the invention of the valve. Finally, I will show that the use of the valve was not clear when it was first made. Fleming actually intended it to be a high-frequency alternating current (AC) measuring instrument for use in the laboratory. Marconi, not Fleming, transformed the valve into a practical detector that was actually used in wireless telegraphy in the field.

Fleming's Recollection of the Invention of the Valve: the Canonical Story and its Problems

In his Friday Lecture at the Royal Institution in 1920, Fleming recalled the invention of the valve by mentioning problems in existing detectors.

> Before 1904 only three kinds of detectors were in practical use in wireless telegraphy—viz. the coherer, or metallic filings detector, the magnetic-wire detector, and the electrolytic detector. ...
>
> The coherer and the electrolytic detectors were both rather troublesome to work with on account of the frequent adjustments required. The magnetic detector was far more satisfactory, and in the form given to it by Senator Marconi is still used. It is not, however, very sensitive, and it requires attention at frequent intervals to wind up the clockwork which derives the moving iron wire band. In or about 1904 many wireless telegraphists were seeking for new and improved detectors.

He then discussed his failed efforts to solve the problem.

> I was anxious to find one which, while more sensitive and less capricious than the coherer, could be used to record the signals by optical means, and also for a personal reason I wished to find one which would appeal to the eye and not the ear only through the telephone. Our electrical instruments for detecting feeble direct or unidirectional currents are vastly more sensitive than any we have for detecting alternating currents. Hence it seems to me that we should gain a great advantage if we could convert the

feeble alternating currents in a wireless aerial into unidirectional currents which could then affect a mirror galvanometer, or the more sensitive Einthoven galvanometer. There were already in existence appliances for effecting this conversion when the alternations or frequency was low—namely, one hundred, or a few hundred per second.

For example, if a plate of aluminium and one of carbon are placed in a solution of sodic phosphate, this electrolytic cell [the Nodon rectifier] permits positive electricity to flow through it from the aluminum to the carbon, but not in the opposite direction. ... But such electrolytic rectifiers, as they are called, are not effective for high frequency current, because the chemical actions on which the rectification depends take time.

In spite of these difficulties, he eventually came to the 'happy moment.'

After trying numerous devices my old experiments on the Edison effect came to mind, and the question arose whether a lamp with incandescent filament and metal collecting plate would not provide what was required even for extra high frequency currents, in virtue of the fact that the thermionic emission would discharge the collecting plate instantly when positively electrified, but not when negatively. ... I found to my delight that my anticipation were correct, and that electric oscillations created in the second coil by induction from the first were rectified or converted into unidirectional gushes of electricity which acted upon and deflected the galvanometer. I therefore named such a lamp with collecting metal plate used for the above purpose, an oscillation valve, because it acts towards electric currents as a valve in a water-pipe acts towards a current of water.[3]

The above recollection clearly shows three consecutive stages which Fleming went through in 1904: 1) the recognition of the trouble with the existing detectors, as well as of the need for a new detector; 2) the recognition of rectification as a new means to detect high-frequency oscillations; 3) finding a method of rectification in his prior research on the Edison effect in bulbs. These three stages provide compelling reasons for Fleming's claim that it was only he who invented the valve. Since they are so convincing, few historians have doubted the story's accuracy. On the demand for stable detectors around 1904, G. Shiers says that Marconi's detectors 'were not satisfactory for regular and dependable service; a new and better detecting device, or signal rectifier, was urgently needed.' Concerning the importance of Fleming's prior research on the Edison effect, Hugh Aitken asserts that 'Fleming's valve was a linear descendent of a device that had no connection with wireless telegraphy at all; this was the famous Edison Effect.'[4]

However, detailed examinations of the technical and corporate factors and contexts under which the valve was invented make some part of Fleming's recollection disputable. First, his story about the existence of a compelling demand for a new detector around 1904 was not entirely convincing, because magnetic and electrolytic detectors, which were more stable and even more sensitive than coherers, were then widely used. Further, one could use a ordinary DC galvanometer with electrolytic detectors.[5] Second, Fleming's research during 1903–1904 had apparently little to do with detectors in wireless telegraphy. His interest in this period largely lay in high-frequency measurement—the measurement of inductance, capacitance, resistance, current, frequency, wavelength, and the number of sparks. Finally, Fleming never wanted to reveal the fact that he was being dismissed from the Marconi Company when he invented the valve in late 1904. In December 1903, his scientific advisorship to Marconi terminated in spite of his wish to retain it. His efforts to recover his connection to the Marconi Company changed Fleming's engineering style. Before 1903, he was distanced from inventing or patenting artifacts, but during the year 1904 he invented and patented the cymometer (a wave-measuring instrument) and the valve. As we will see, these two artifacts—the valve, in particular—helped him to resume his connection to the Marconi company in May 1905.

This paper will propose a different history of the valve. The central thread of my history is the interaction between Fleming and local resources—material, social, conceptual, and linguistic—available to him. These resources constrained and promoted his laboratory practice. Fleming's goal was shaped and materialized while mobilizing, changing, and combining these resources.[6] These resources, his laboratory practice, and his goal co-shaped each other. To show this, I will start with Fleming's work on high-frequency measurement in 1902–1904.

Measuring High-Frequency Alternating Current with Rectification

At the turn of the century, there were two different methods for measuring high-frequency alternating current. The first method was to use the hot-wire amperemeter. It utilized the elongation of a metallic wire when heated by feeble high-frequency alternating current. The second method was to use sensitive AC dynamometers. Ordinary hot-wire amperemeters and dynamometers were not however sensitive enough to measure feeble high-frequency current that one had to measure in wireless telegraphy. Therefore, Fleming devised a third method, in which the effect of the magnetization of iron by electromagnetic waves was exploited.

The magnetization of a piece of demagnetized iron, as well as the demagnetization of a piece of magnetized iron, by electromagnetic waves was discovered by E. Rutherford in 1896. Based upon this, Marconi invented a practical magnetic detector in 1902. In designing his magnetic detector, Marconi used a telephone, not a galvanometer, because ordinary DC galvanometers could not detect high-frequency AC signals. In December 1902, Fleming and his assistant, A. Blok, decided to investigate how the Rutherford effect might be made to work with a galvanometer. The operational principle they tried to utilize was simple. Suppose, they thought, that electromagnetic waves were allowed only to demagnetize the iron, which was then magnetized again by some other means. Then, the effect of electromagnetic waves on the magnetic detector could be exhibited by an ordinary DC galvanometer. If electromagnetic waves were made to act only in one-way action (either to magnetize or to demagnetize), then this effect could be converted into electrical signals to be detected with a DC galvanometer. After much trial and error, Fleming and Blok constructed a workable device. Fleming described it before the Royal Society in March 1903.[7]

The measuring instrument is shown in Figure 3. In this, *bb′* is a large outer bobbin and inside it lie wire bundles upon which an ordinary magnetizing coil *ad′* and a demagnetizing coil *cc′* are wound (shown in the below). Magnetizing coil *ad′* is connected with battery *P* and demagnetizing coil *cc′* is connected with the receiving aerial and the earth. One end of outer bobbin *bb′* is connected with the galvanometer *G*. The commutator *C*, shafted to a 500 rpm motor, consists of five disks numbered *1*, *2*, *3*, *4*, and *5*. Each disk has a fan-shaped sector made of brass, which occupies a certain angle. The angle of the brass sector of disk *1* is 95 degrees, *2* and *3* is 135 degrees, *4* is 140 degrees and *5* is 360 degrees. Disk *1* and *5* are connected to *a* and *d′*, *2* and *3* to *b* and *b′*, and *4* and *5* to the galvanometer *G*.

Commutator operation is as follows. When the commutator starts rotating, the battery magnetizes the iron bundle by the action of disk *1* and *5*. This does not affect the galvanometer needle because during this time the galvanometer circuit is disconnected by disk *4*. For a short time after the magnetizing current is stopped, the secondary bobbin is solely connected with the galvanometer circuit, because the brass sectors of disk *2* and *3* (135 degrees) are larger than that of disk *1* (95 degrees). After it was disconnected, a connection was made with the galvanometer circuit by disk *4* and *5*. Suppose that electric waves strike the antenna and then oscillations pass

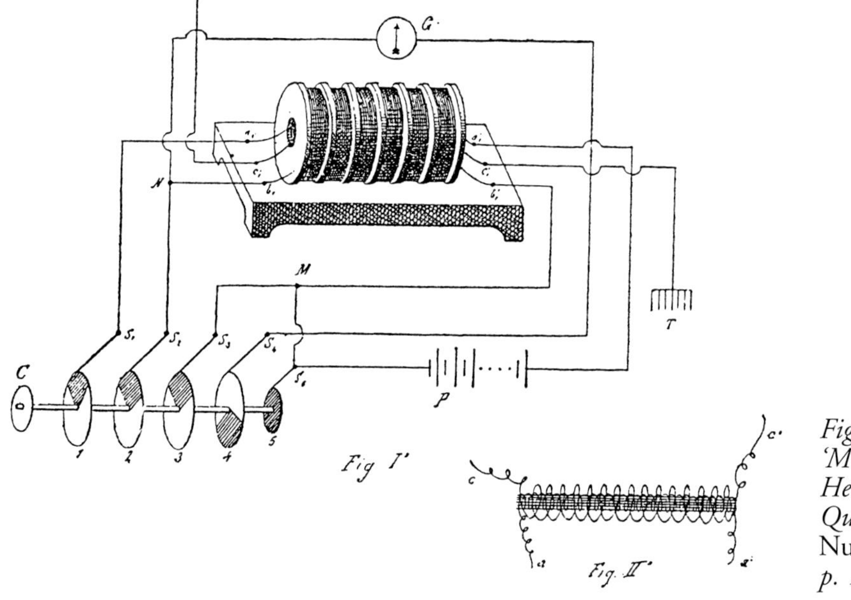

Figure 3. Fleming's 'Magnetic Detector for Hertzian Waves adapted for Quantitative Work.' Source: Nuovo Cimento *9 (1905), p. 108.*

through *cc′* to the earth. This demagnetizes the iron bundle that has already been magnetized, which causes a change in the magnetic field in the interior of the bobbin, and produces an electric current in the bobbin wires *bb′*. The current is detected by the deflection of the galvanometer needle. Suppose that one rotation of the commutator occurs for a very short time (about 1/10 second), during which the galvanometer needle does not return to its normal place, then the effect accumulates and the deflection of the needle becomes steady until oscillation continues. This was what Fleming wanted:

> Since the interrupter discs are rotating very rapidly, if the electrical oscillation continues, these intermittent electromotive impulses *produce the effect of a continuous current in the galvanometer circuit*, resulting in a steady deflection, which is proportional to the demagnetizing force being applied to the iron, other things remaining equal.[8] [emphasis added]

In short, the device showed an effect of rectifying high-frequency alternating current. The instrument 'converted' it into a direct current that was measured by an ordinary galvanometer.

What were the uses of this device? Fleming listed several primary uses:

> By means of such an arrangement it is possible to verify the law according to which variation falls off with distance. *The instrument can be employed also as a telegraphic receiving instrument, but its chief use will be for comparing together the wave-making power of different radiators.* ... This detector serves, for instance, to show in a very marked manner the great effect of slight difference in the surface of the spark balls. ... Such an instrument will probably be found of great use in connection with the design of radiators and transmitters for Hertzian wave wireless telegraphy. ... Similarly, the instrument promises to be of considerable use in the investigation of the transparency or opacity of various substances to Hertzian waves, not merely qualitatively, but in the determination of a coefficient of absorption.[9] [emphasis added]

It is important to notice here that the device was not merely a detector in its usual sense; it was rather a laboratory instrument, a kind of high-frequency AC galvanometer. Used as a detector, it was very cumbersome and much less efficient than most detectors available at that time. A month after, Fleming again stressed the demanding need of a meter that could give a measure of the energy of waves quantitatively. 'It is only by the possession of such an instrument' he said, 'that we can hope to study properly the sending powers of various transmitters, or the efficiency of different forms of aerial, or devices by which the wave is produced.'[10] Remember that the measurement of the powers of transmitters was Fleming's primary goal.

However, the instrument was not widely used. Besides Fleming, as far as my knowledge goes, it was only used by an Italian scientist, V. Buscemi, in examining absorptions of electromagnetic waves in various dielectrics.[11] The reason for this neglect was due to its lack of stability. It turned out to be very difficult to calibrate this device because of the complex commutator actions. After this failure, Fleming devised sensitive hot-wire amperemeters, but it was not sensitive enough to measure feeble current less than 5 milliamperes, a current which wireless engineers had to frequently measure in the laboratory and the field.[12]

In December 1903, when Fleming was still struggling with this measurement problem, he was informed that his scientific advisorship to the Marconi Company would not be renewed. Between 1899 and 1903, Marconi needed Fleming as an expert on patent and as a credible witness of Marconi's secret demonstration. Besides, Fleming played the role of a bridge between Marconi and the British scientific and engineering communities. Fleming also helped Marconi to design a high-power transmitting station at Poldhu for the first transatlantic experiment in 1901. The transatlantic transmission of wireless signals was a huge success, but Fleming was deeply hurt by Marconi's apparent attempt to monopolize credit for the success. Marconi was then much disturbed by Fleming's request to share it. After that, a subtle tension existed between them. The Maskelyne affair of June 1903, in which Nevil Maskelyne, Marconi's adversary, ruined Fleming's public demonstration of Marconi's new syntonic system by sending derogatory messages for interference, hurt much of Fleming's credibility.[13] His advisorship was scheduled to end in December of 1903. The Marconi Company did not renew it.

Fleming first approached to the Marconi Company with his new ball discharger, but it did not move Marconi who had already thought that Fleming's ball discharger was no more than a gadget. Fleming sensed the urgent need to invent something more useful. In the summer of 1904, he invented the cymometer (a convenient wave-length measuring instrument), and a few months later, the valve. These two artifacts provided Fleming with what he wanted: a revival of the connection to the Marconi Company.

The Invention of the Valve

Fleming invented and applied for a patent on the cymometer in June 1904.[14] After this he returned to the measurement of current. Since specially constructed AC amperemeters and dynamometers proved to have been unreliable for high-frequency alternating current, he returned to the original method of effectively rectifying it. As before, he borrowed technologies from power engineering. Converting AC into DC was an important subject in power engineering. Several rectifiers such as rotary converters and electrolytic cells had been widely used. Obviously, the rotary converter, a huge machinery used in power engineering to convert between AC and DC, was not a candidate.[15] The second candidate, some electrolytic cells, looked more promising. Fortunately for Fleming—and this is what I want to argue—he found what he had been looking for. The *Electrician* of 14 October (1904) published a paper by A. Nodon, a French electrician, entitled 'Electrolytic Rectifier: An Experimental Research.' This article described the characteristics and the efficiency of Nodon's aluminium cell rectifiers, which consisted of two metals (one usually aluminium) dipped in an electrolyte (carbonate of ammonium, for example). With AC of 42–82 Hz, Nodon obtained 65–75% rectification. In this article, Nodon described his rectifier as electrolytic 'valve'—which suggests the origin of the name of Fleming's valve—and used the term 'valve effect' and the 'action of the valve.' Fleming constructed Nodon's valve to replicate the rectifying effect with high frequency oscillations, but after many experiments he eventually found that the Nodon valve was not effective for high-frequency alternating current.[16]

The Nodon valve illuminates some interesting points. First, we can understand why Fleming later repeated that 'electrolytic rectifiers' were inefficient for the rectification of high-frequency oscillations. Lee de Forest once criticized Fleming over this point, because most electrolytic detectors in fact rectified high-frequency oscillations. But what Fleming meant by the electrolytic rectifier was not ordinary electrolytic detectors used in wireless telegraphy, but the Nodon electrolytic rectifier designed for power engineering. Secondly, Fleming's attention to the Nodon valve supports my argument that Fleming was less concerned with the new detectors than with a metrical instrument for high-frequency current measurement. If Fleming had been really interested in detectors, he would have experimented on existing electrolytic detectors, rather than Nodon rectifiers which had nothing to do with wireless telegraphy. Finally, although Fleming later mentioned that he invented the valve in October 1904, the Nodon rectifier, noticed by Fleming in mid-October, suggests that the discovery might have been later. In fact, in a short letter to the *Electrician* in 1906, Fleming said that '*in November, 1904*, [I] discovered that this unilateral conductivity [in the lamp] held good for high frequency current, which is not the case for electrolytic rectifiers.'[17] [emphasis added]

After his failure with the Nodon valve, Fleming pondered other means to create the valve effect (rectification) for high-frequency oscillations. What he needed was a device that allowed only one-way flow of currents. This reminded him of the unilateral conductivity in the small space inside the bulb, a phenomenon which he had investigated a few times in the 1880s and 1890s. He asked his assistant, G. B. Dyke, to connect a circuit as instructed, and 'took out of a cupboard one of [his] old experimental bulbs.' It immediately proved workable (see Figure 4 and Figure 5).[18]

Why at this moment did Fleming come up with the idea of using the lamp for rectification? I have suggested that an 'invention imperative' pressed Fleming to design something useful. Nodon's paper on electrolytic 'valve' was published in October 1904, which provided Fleming with conceptual and linguistic resources such as the term valve and the idea of the valve effect. In addition to these, the electron theory that Fleming adopted around 1900

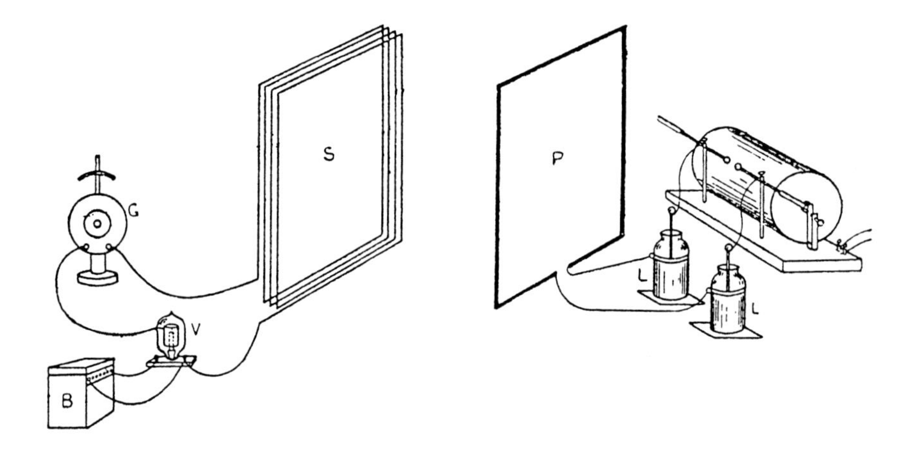

Figure 4. Fleming's First Test on Rectifying Characteristic of the Valve.
Source: Proceedings of the Physical Society *20 (1906), p. 179.*

enabled Fleming to reinterpret the mechanism of the Edison effect. In the 1880s and 1890s, unilateral conductivity in the Edison effect was explained by Fleming in terms of the emission of negatively charged carbon molecules from the filament of a lamp to an inserted third electrode. A battery attached to the third electrode functioned as a discharger of negative charges accumulated on it, and a galvanometer measured the rate of this discharge. In 1904, Fleming was able to reinterpret the mechanism of the Edison effect in a radically different way. The battery attached to the third circuit functioned as a

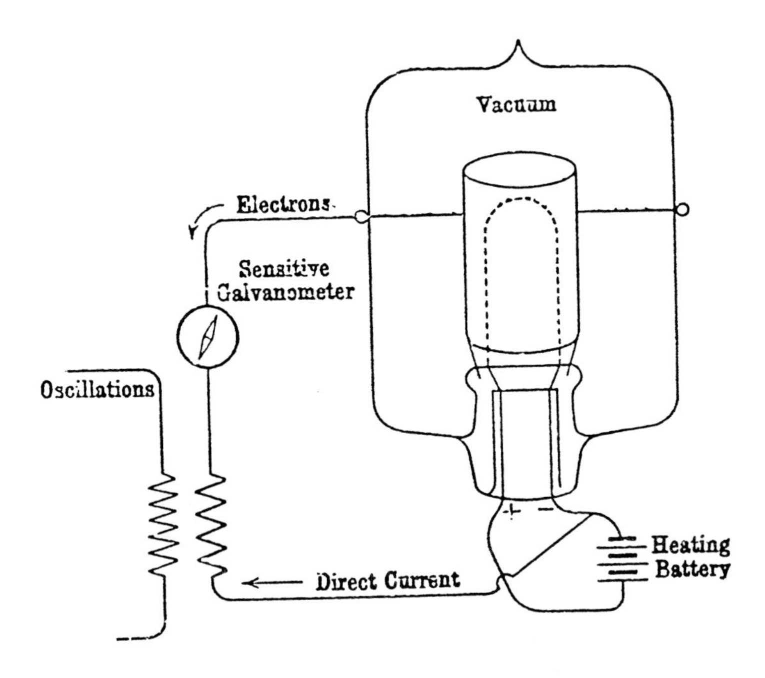

Figure 5. Rectification by the Valve. This is one of the earliest printed circuits on the valve. Source: Electrician *54 (31 March 1905), p. 967.*

source of electrons which were pushed into the vacuum, and the galvanometer measured the rate of this current of electrons. Unilateral conductivity was transformed into a valve action.[19]

Constructing the Uses of the Valve

After his first successful experiment, Fleming filed a provisional specification of the British patent, No. 24,850, on 16 November 1904. The title of the patent was 'Improvements in Instruments for Detecting and Measuring Alternating Electric Currents.' We can confirm the purpose of his invention in this provisional specification. Nowhere did he mention troubles in existing detectors, nor did he allude to putative demand for new ones. All that he mentioned was his former research on the measurement of high frequency alternating current. After this, he explained his purpose: 'The object of my invention is to provide a means by which an ordinary galvanometer can be used to detect and measure alternating electric currents and especially high frequency currents commonly known as electric oscillation.' Then, he mentioned two uses of the valve: 1) 'the device is especially applicable to the detection and measurement by an ordinary galvanometer of high frequency current or electric oscillations, where any form of mechanical or electrolytic rectifier is useless'; 2) the device 'and a galvanometer *may be used* as a receiving instrument in wireless telegraphy.' The first objective stressed its use (with an ordinary galvanometer) as a current measurer in the laboratory, and the second one indicated its possible use as a receiver in wireless telegraphy.[20]

Between December 1904 and January 1905, Fleming conducted a series of experiments with his single valve—the only valve working satisfactorily—to determine its conductivity at different electrode voltages. The experiments showed that the current-voltage (I-V) characteristics of the valve were neither linear nor regular. This made its rectifying power highly unpredictable. It undermined Fleming's initial hope to have produced a metering device. What about the valve as a detector, then? As a detector, 'the arrangement [a valve with a galvanometer in Figure 4], although not as sensitive as a coherer or magnetic detector, is much more simple to use.' Manageability was a merit and bad sensitivity was a weakness. But another merit, such as detecting 'a change in the wave-making power or uniformity of operation of transmitting arrangement,' would make up for this disadvantage.[21] Though not entirely linear, the valve had an instrumental or a metrical character that was lacking in other detectors. In sum, the valve had *potential* merit as a '*metrical detector*,' not simply as a metre or a detector.

But its obvious and immediate use lay elsewhere. The valve seemed useful in rebuilding his connection to Marconi and the Marconi Company. Having filed a patent for his valve, Fleming wrote to Marconi. In this undated letter (written in late November), he first mentioned his invention of the wavelength measuring instrument (the cymometer), and then remarked:

I have found a method of rectifying electrical oscillation, that is, making the flow of electricity all in the same direction, so that I can detect then with an ordinary mirror galvanometer. I have been receiving signals on an aerial with nothing but a mirror galvanometer and my device, but at present only on a laboratory scale. This opens up a wide field for work, as I can now measure exactly the effect of the transmitter.

Then, Fleming assured Marconi that 'I have not mentioned this to anyone yet as it may become very useful.'[22]

Fleming's new inventions impressed Marconi who expressed his wish to perform some experiments with Fleming's 'electrical oscillation current rectifier.' Fleming proposed setting up his apparatus at Marconi's factory at Chelmsford, where Marconi 'might come to see it at some convenient time.' In his Friday Lecture on 3 March 1905, Marconi exhibited the cymometer and the valve. After the lecture, Marconi asked Fleming if he could borrow the valve for more experiments.[23] This was an opportunity for Fleming to firmly re-establish his connection with Marconi, one that Fleming took care not to miss. He immediately replied to Marconi:

The only valve I have in my possession which works well is the one I lent you for your lecture, and with which all the work for my Royal Society paper was done. I do not wish to part with this valve as I shall have then no means of making comparison measurements. I will endeavour to get a couple of good valves made as soon as possible and sent over to you. Meanwhile I should like to draw your attention to the matter of my agreement with the Company about which I spoke to you when we last meet. In your absence I trust the matter will not be prolonged as it will be to the interest of the Company that it should be reestablished.[24]

A few months later, in May 1905, the Marconi Company re-appointed Fleming as scientific advisor starting on the first of that same month. A Memorandum of Agreement was signed between the two directors of the Company and Fleming. The agreement contained four terms, the third of which specified that Fleming's inventions made between 1 December 1903 and 1 May 1905 would be put under the same agreement as if he had been a scientific advisor during that time.

In particular the said John Ambrose Fleming shall from time to time and at all times during the said period [1 May 1905–30 April 1908] communicate to the Company free of all charge and expense all improvements discoveries and inventions which he may hereafter make or become acquainted with in connection with the Company's business or any part thereof and shall also at the request and cost of the Company but free of charge use his best endeavour to obtain letters patent and/or the foreign equivalents thereof in such countries as the Company may desire in respect of any such improvements discoveries and inventions (*whether the same may have been made or brought to his knowledge heretofore since the expiration of the first above mentioned period of three years from the first day of 1 December, 1900, or may be made or brought to his knowledge hereafter during the period of this agreement*) ...[25]

Though the patent on the valve was issued in the single name of Fleming, the Marconi Company now had rights to it.

For the following few years until around 1907, Fleming, not Marconi, ordered the fabrication of the valve at the workshop of the Ediswan Company in Ponders End. Charles Gimingham of the Ediswan Company, a skilled artisan of lamp construction and Fleming's life-long friend, made them under Fleming's instruction. Whenever Marconi and the Marconi Company asked Fleming for valves, Fleming ordered them and Fleming's assistant, G. B. Dyke, took them to the Marconi Company. But, even during this early period, the *use* of the valve was not under Fleming's control. In late 1906, one Italian professor asked Fleming whether he could obtain the valve for scientific experiments. Thinking it possible, Fleming told Marconi about this, saying that 'even if you wish to keep the exclusive use of the valve for wireless telegraphy, it might still be possible to supply it for such other scientific uses as do not conflict with your work.'[26] Marconi, however, refused to hand over the valves to others, on the basis of his current experiments on the valve as a long-distance wireless detector. In 1907, however, Marconi decided to hand over several valves to the Italian Government, and Fleming was asked to give instruction to the Italians.

Marconi's experiment, not Fleming's, transformed the valve into a sensitive detector for wireless telegraphy. In mid-1905, while experimenting with the device, Marconi found that it became much more sensitive when he replaced the galvanometer with a telephone. In 1906–1907, Marconi increased its sensitivity by connecting a telephone inductively to the valve circuit using his oscillation transformer or 'jigger.' Only with this modification did the valve become 'one of the best long distance receiver yet made, under some conditions better even than the magnetic detector.' But it was still far from a manageable device. It developed problems when used with Marconi's new 'continuous waves' (which were actually quasi-continuous waves with reduced damping). In his station at Poole, Marconi could receive clear signals sent from Poldhu with other receivers, but not with the valve, when the continuous wave was used.[27]

In 1908, Fleming devised a tungsten valve and a new circuit for connecting the valve to the receiver. As in November of 1904, Fleming wrote a letter to Marconi, saying that 'I am anxious this should not be yet known to anybody but yourself.' Marconi, however, did not welcome the new modifications. After some experiments, he informed Fleming that 'Tungsten valves are not more sensitive than the best carbon valves ... [and] that your new arrangement of circuit does not give as good results as the standard circuit which ... is strictly in accordance with the description given in my [Marconi's] patent for the mode of using your valve as a receiver.' After that time, Fleming made no further important contribution to the development of the valve.[28]

During this early period, Fleming apparently was not concerned with the use of his valve by others. In October, 1906, Fleming told Marconi:

[The patent on the valve] is not by any means a strong patent and if we refuse to supply the valve for all purposes, people may import the similar device of Wehnelt from Germany or perhaps make it for themselves. I can hardly believe that the patent is worth fighting. ... Personally I have no interest in the matter; I have already any little scientific credit there may be for the invention.[29]

Just after this, however, the development of the valve took an unforeseen direction. Lee de Forest in America announced his invention of the audion in October 1906. Basically, de Forest inserted a grid (applied to a separate potential difference) in the vacuous space between the plate and the filament, by means of which the current flow in the space was controlled. With this improvement, de Forest argued that the audion showed not only rectification, but also amplification, of received signals. He tried to undermine Fleming's earlier contribution as much as he could. As to its scientific side, de Forest stressed that Fleming's work on unilateral conductivity in the Edison effect had already been anticipated by Elster and Geitel. De Forest argued that the use of Fleming's valve was confined to 'quantitative measurements over short distances.' He even remarked:

The value of such a device [Fleming's valve] as a wireless telegraph receiver is nil. Exceedingly powerful signals are required to affect this galvanometer at all. Put a telephone in the circuit, and if the signals are sufficiently intense you can detect the pulsating currents in the telephone. But it is not practicable in commercial wireless telegraphy.[30]

Even though Fleming did not read de Forest's paper in full, the abstract published in the *Electrician* was enough to infuriate him. Fleming immediately countered that 'the actual construction of the apparatus [the Audion] is the same [as mine]' and that his valve had been actually used as a sensitive receiver in wireless telegraphy. In his reply, de Forest again argued that the real genesis of his audion, as well as Fleming's valve, was Elster and Geitel's research in the 1880s. He then compared the valve to 'a laboratory curiosity' and the audion to 'an astonishingly efficient wireless receiver employing the same medium, but operating on a principle different in kind.'[31] Fleming's credibility was threatened, which was unbearable to him. This brought him into an alliance of mutual interest with Marconi, since de Forest was one of Marconi's chief competitors. Fleming thus urged Marconi:

It is extremely important that de Forest in America should not be allowed to appropriate all methods of using the glow lamp detector ... I want our Company to have all the commercial advantage possible but I am anxious that de Forest shall not deprive me of the scientific credit of the valve invention as he is anxious to do ...[32]

This was the beginning of Fleming's life-long animosity against de Forest. They again crossed swords in 1913 through the column of the *Electrician.* But the controversy was not to be resolved in engineering journals; it was resolved in court. In 1917, Fleming began his legal battle with de Forest in the court of United States of America. Four years later, in 1921, Fleming's priority in the utilization of the two electrode lamps in wireless telegraphy was supported by the U. S. court.[33]

De Forest's audion apparently forced Fleming to ponder the originality of his invention. Fleming knew that originality did not lie in the device itself—a lamp with one more electrode—because it had been made by Edison, as well as by Elster and Geitel. Rectification of high-frequency alternating current was insufficient, because it could be regarded as 'a laboratory curiosity.' De Forest persistently argued that only his audion suited to practical detection. Fleming for his part repeatedly emphasized that what he had invented was a new detector for wireless telegraphy, whereas what de Forest had done was simply to add one more plate to his valve. This way of thinking seemed to be firmly implanted in his mind after the patent litigation against de Forest. Fleming first recalled how he invented his valve in 1920, when he was in the middle of the battle with de Forest. The same story was then repeatedly told. It was reiterated by engineers and historians of later generations.

Conclusion

Fleming's valve became more famous after the 'triode revolution' in the mid-1910s.[34] The triode, or the audion, began to be used not only as an amplifier but also as an oscillator for continuous waves. As the audion became more essential to radio engineering, the valve—as audion's predecessor—was increasingly highlighted. However, Fleming was not always happy with the changed situation. In a sense, Fleming became increasingly frustrated. In 1918, the Marconi Company applied for an extension of Fleming's patent, but it was dismissed on the grounds that the Company had earned sufficient profit by the sale of Fleming's valve. However, 'as the original inventor of it,' Fleming later complained 'I have never received a single penny of reward for it.' In addition, in Fleming's view, the Company hurt Fleming's credibility by dissociating Fleming's name from his invention. Only after he resigned the scientific advisorship to the Marconi Company, he confessed that 'one firm has sold valves for many years made exactly in accordance with my patent specification, but which they advertise and mark "Marconi Valves."'[35] This 'injustice of some present-day commercial practice' that he lamented was the price he paid for resuming his connection with the Marconi Company in 1905.

Fleming's recollection of the invention of the valve can be characterized by the continuity of the device to his previous work (the Edison effect), as well as by the radical discontinuity between his and others' work. He stressed his 'scientific' research on the Edison effect that he had conducted in the 1880s and 1890s, the importance of his role as scientific advisor to Marconi, and the demand for new detectors around 1904, as well as the valve's straightforward use as a sensitive detector when it was invented. However, I have revealed complexities and contingencies behind this neat story of Fleming. The invention turns out to have more to do with local and temporal resources specifically available to Fleming in the fall of 1904, than his long-term research on the Edison effect or the putative high demand for new detectors for wireless telegraphy. I argued that the combination of heterogeneous factors such as Fleming's previous research on the measurement of high-frequency alternating current by rectification, Fleming's dismissal from the Marconi Company at the end of 1903, Nodon's electrolytic valve publicized by the *Electrician* in October 1904, and the electron theory formed technical, material, conceptual, and linguistic resources that Fleming mobilized in a timely manner.

I also tried to show that the use of the valve was not clearly defined in 1904. Fleming intended it to be a measuring instrument, a meter. Its irregular and nonlinear current-voltage characteristics made it very difficult to use the device as a meter. As a detector, it was less sensitive than other detectors. But it had a social use. It was successfully exploited to revive Fleming's severed connection to the Marconi Company. As a detector, it remained auxiliary until Marconi connected a telephone and a jigger to the valve circuit. I have also shown that Fleming became much concerned about his credit as the inventor of the valve when de Forest depreciated Fleming's originality. The 'canonical' story about the invention of the valve was first remembered by Fleming when he was deeply engaged in a lawsuit with de Forest.

I have tried to criticize this 'canonical' story by discovering various factors which I believe shape a more appropriate and compelling context for Fleming's invention. I have also located Fleming's canonical story in its own context—the context of his battle with de Forest and the subsequent patent litigation. The usefulness of the device and its history have been constructed and consolidated into the form that we now know today in a very similar fashion. Inventors not only invent a novel artifact; they also invent a novel history for it.

Acknowledgements

I thank the Library of the University College London for permitting me to quote the archives in the Fleming Collection, and the Science Museum for providing me with the two pictures of Fleming's special lamps used in 1889. I am grateful to an anonymous referee for helpful suggestions.

Notes

1. For the interrelatedness between recalled narratives of an invention and patent litigation, see Sungook Hong, 'Marconi and the Maxwellians: the Origins of Wireless Telegraphy Revisited,' *Technology and Culture* 35 (1994), 717–749. The constructed nature of the history of scientific discoveries written by the actors involved in them is well discussed by Simon Schaffer, 'Scientific Discoveries and the End of Natural Philosophy,' *Social Studies of Science* 16 (1986), 387–420. The role of a collective memory for the scientific community is examined by Pnina Abir-Am, 'A Historical Ethnography of a Scientific Anniversary in Molecular Biology: The First Protein X-ray Photograph,' *Social Epistemology* 6 (1992), 323–354.
2. Another paper of mine focuses on the transformation from the Edison effect to the valve. See Sungook Hong, 'From Effect to Artifact (II): The Case of the Thermionic Valve,' *Physis* 33 (1996), 85–124.
3. J. A. Fleming, 'The Thermionic Valve in Wireless Telegraphy and Telephony' (Friday Evening Lecture on 21 May 1920), *Proceedings of the Royal Institution* 23 (1920), 167–168. See also J. A. Fleming, 'How I Put Electrons to Work in the Radio Bottle,' *Popular Radio* 3 (1923), 175–182.
4. G. Shiers, 'The First Electron Tube,' *Scientific American* 220 (March 1969), 109; Hugh G. J. Aitken, *The Continuous Wave: Technology and American Radio, 1900–1932* (Princeton, 1985), p. 205.
5. For magnetic and electrolytic detectors, see, respectively, T. H. O'Dell, 'Marconi's Magnetic Detector: Twentieth Century Technique Despite Nineteenth Century Normal Science?' *Physis* 25 (1983), 525–548 and V. J. Phillips, *Early Radio Wave Detectors* (London, 1980), pp. 65–84.
6. For some general considerations of experimental culture in the laboratory, see Jed Buchwald, *The Creation of Scientific Effects: Heinrich Hertz and Electric Waves* (Chicago, 1994); Ian Hacking, 'The Self-Vindication of the Laboratory Science,' in Andy Pickering (ed.), *Science as Practice and Culture*, (Chicago, 1992), pp. 29–64; Andy Pickering, 'The Mangle of Practice: Agency and Emergence in the Sociology of Science,' *American Journal of Sociology* 99 (1993), 559–589.
7. J. A. Fleming, 'A Note on a Form of Magnetic Detector for Hertzian Waves adapted for Quantitative Work,' *Proceedings of the Royal Society* 71 (1903), 398–401.
8. J. A. Fleming, 'A Note on a Form of Magnetic Detector,' p. 400. Twelve years later, Fleming modified it as follows:

   > Since the interrupter discs are rotated very rapidly, if the oscillations continue, this intermittent electromotive force produce *a practically steady current* through the galvanometer which is proportional to the demagnetizing force being applied to them.
   >
   > J. A. Fleming, *An Elementary Manual of Radiotelegraphy and Radiotelephony* (London, 1916), p. 205.
9. J. A. Fleming, 'A Note on a Form of Magnetic Detector,' pp. 400–401.
10. J. A. Fleming, 'Hertzian Wave Telegraphy' (Cantor Lecture given on 16 March 1903), *Journal of the Society of Arts* 51 (1902–03), 762.
11. V. Buscemi, 'Trasparenza dei Liquidi per le onde Hertziane,' *Nuovo Cimento* 9 (1905), 105–112. Figure 3 comes from this paper.
12. Fleming's hot-wire amperemeter was able to measure 10 milliamperes within an accuracy of 2–3 percent and it could detect 5 milliamperes, but did not measure below 5 milliamperes, which was its serious weakness. See J. A. Fleming, 'On a Hot-Wire Ammeter for the Measurement of Very Small Alternating Currents' (read at the Physical Society on 25 March 1904), *Proceedings of the Physical Society* 19 (1904), 173–184.
13. For this, see Sungook Hong, 'Styles and Credit in Early Radio Engineering: Fleming and Marconi on the First Transatlantic Wireless Telegraphy,' *Annals of Science* 53 (1996): 431–465; 'Syntony and Credibility: John Ambrose Fleming, Guglielmo Marconi, and the Maskelyne Affair,' *Archimedes* 1 (1996), 157–176.
14. For a useful description of various rectifiers, see P. Rosling, 'The Rectification of Alternating Currents,' *Electrician* 58 (1906), 677–679. The rotary converter was a 'gateway technology' which enabled the coupling of different networks of AC and DC systems. For this, see Thomas Hughes, *Networks of Power: Electrification in Western Society, 1880–1930* (Baltimore and London, 1983), pp. 120–121.
15. For a useful description of the rectifiers, see P. Rosling, 'The Rectification of Alternating Currents,' *Electrician* 58 (1906), 677–679.
16. A. Nodon, 'Electrolytic Rectifier: An Experimental Research' (read at the St. Louis International Congress), *Electrician* 53 (14 October 1904), 1037–1039. On the description of Fleming's trial with the Nodon valve, see J. A. Fleming, 'On the Conversion of Electric Oscillation into Continuous Currents by means of Vacuum Valve,' *Proceedings of the Royal Society* 74 (1905), 476.

17. For the dispute between Fleming and de Forest over electrolytic detectors, see Fleming's and de Forest's letters to the Editor of the *Electrician* in 1906. J. A. Fleming, 'Oscillation Valve or Audion,' *Electrician* 58 (30 November 1906), 263; L. de Forest, 'Oscillation Valve or Audion,' *Electrician* 58 (29 December 1906), 425; J. A. Fleming, 'Oscillation Valve or Audion,' *Electrician* 58 (4 January 1907), 464. The quotation is from Fleming's 26 November letter. It was the first published identification of the date, but after 1920 he said that he had invented it someday in October 1904. For this 'October-invention' story, see J. A. Fleming, *Memories of a Scientific Life* (London, 1934), p. 141.
18. J. A. Fleming, *Memories*, p. 141. The first circuit is described in J. A. Fleming, 'The Construction and Use of Oscillation Valves for Rectifying High-Frequency Electric Currents,' *Proceedings of the Physical Society* 20 (1906), 177–185.
19. The role of the electron theory for the invention of the valve is described in detail in Sungook Hong, 'The Electron Theory and Wireless Telegraphy,' a paper read at the workshop on the 'Electron and the Birth of Microphysics' (Dibner Institute, MIT, 10 May 1997). For Fleming, as for J. Larmor, the electron was not a solid particle like a billiard ball. It was instead a kink, or knot, of the ether strain. Fleming detailed his electron theory in his Friday Lecture, 'The Electronic Theory of Electricity' (Friday Lecture given on 30 May 1902), *Proceedings of the Royal Institution* 17 (1902), 163–181.
20. J. A. Fleming, 'Improvements in Instruments for Detecting and Measuring Alternating Electric Currents,' Provisional specification, British patent No. 24,850, 16 November 1904.
21. Fleming, 'On the Conversion of Electric Oscillation,' p. 480.
22. Fleming to Marconi, n.d., (circa 30 Nov. 1904), Marconi Company Archives [hereafter abbreviated MCA], Chelmsford. G. Shiers, who argues that Fleming invented his valve because of the urgent need for new detectors, presented this letter of Fleming's to Marconi as a piece of supporting evidence. However, nowhere in this letter did Fleming mention that he had found a new detector. The phrase that 'I can now measure exactly the effect of the transmitter,' in fact, supports my point. See Shiers, footnote 4, p. 111. Aitken, footnote 4, p. 211, also quotes this letter from Shiers, but Aitken seems to have noticed that Fleming invented it as a meter, not as a detector (on p. 212).
23. Marconi to Fleming, 14 Feb. 1905, University College London [hereafter abbreviated UCL] MS Add 122/1, Fleming Collection. Fleming to Marconi, 15 Feb. 1905, MCA, Chelmsford. G. Marconi, 'Recent Advances in Wireless Telegraphy' (Friday Lecture on 3 March 1905), *Proceedings of the Royal Institution* 18 (1905), 31–45.
24. Fleming to Marconi, 24 March 1905, MCA, Chelmsford.
25. 'Memorandum of Agreement between the Marconi Company and Fleming' 26 May 1905, MCA, Chelmsford.
26. Fleming to Marconi, 9 Oct. 1906, MCA, Chelmsford.
27. (A copy of) Fleming to A. R. M. Simkins (Work Manager of Marconi Company), 17 Jan. 1907, UCL MS Add 122/48, Fleming Collection; Marconi to Fleming, 14 April 1907, UCL MS Add 122/48, Fleming Collection; (A copy of) Fleming to Marconi, 15 April 1907, UCL MS Add 122/48, Fleming Collection.
28. (A copy of) Fleming to Marconi, 8 Sep. 1908, UCL MS Add 122/48, Fleming Collection; Marconi to Fleming, 8 Jan. 1909, UCL MS Add 122/48, Fleming Collection.
29. Fleming to Marconi, 9 Oct. 1906, MCA, Chelmsford.
30. Lee de Forest, 'The Audion: A New Receiver for Wireless Telegraphy,' *Transactions of the American Institute of Electrical Engineers* 25 (1906), 748, 775. Its abstract was published in *Electrician* 56 (1906), 216–218.
31. See the exchanges between Fleming and de Forest cited in note 17.
32. (A copy of) Fleming to Marconi, 14 Jan. 1907, UCL MS Add 122/48, Fleming Collection.
33. For the story of the patent litigation over the vacuum tube see G. W. O. Howe, *The Genesis of the Thermionic Valve in Thermionic Valves 1904–1954: The First Fifty Years* (London, 1955), pp. 15–16.
34. For the 'triode revolution' refer to Sungook Hong, 'The Triode Revolution: A New Interpretation' (Colloquium at the Institute for the History and Philosophy of Science and Technology, University of Toronto, 25 November 1998).
35. Fleming, *Memories*, p. 147.

Alan Q. Morton

The Electron Made Public: The Exhibition of Pure Science in the British Empire Exhibition, 1924–5

In 1997 the centenary of the discovery of the electron was celebrated in numerous lectures, exhibitions, and publications. These events commemorated the outcome of an experiment described in a standard physics text in the following terms.

'In 1897 J. J. Thomson, working at the Cavendish Laboratory in Cambridge, measured the ratio of the charge e of the electron to its mass m by observing its deflection in combined electric and magnetic fields. The discovery of the electron is usually said to date from this historic experiment. ...'[1]

For the physics community the celebrations commemorated an important episode the history of their discipline. For not only did Thomson and his colleagues represent the first generation of professional physicists, their work on X-rays, radioactivity, and the electron led to remarkable breakthroughs in understanding the internal working of atoms. In turn understanding the atom provided a new basis for research in many areas of physics. This new knowledge about the atom, quantum mechanics, and its practical outcomes, from electronic devices to nuclear weapons, changed the course of human history.[2] Consequently, the celebrations marking the discovery of the electron were opportunities for physicists to draw the attention of a wide audience to the benefits (but usually not to the disbenefits), of their work. In fact, accounts of the electron centenary reached a wide audience through mass media such as radio and television, and their newer electronic incarnations, whose operation depends on manipulating electrons.

Why should J. J. Thomson's experiment be singled out? To answer this question we shall consider the first public presentation of the account that Thomson discovered the electron in the course of a particular experiment in 1897. The occasion was the British Empire Exhibition held at Wembley, near London, in 1924–5 when the Exhibition of Pure Science displayed a cathode ray tube Thomson had used in his 1897 experiments. It was described as the

'Apparatus by which the existence of electrons was detected and their mass and velocity measured.'[3]

While this claim was in accord with the views of professional scientists, it was also shaped by the circumstances of public exhibition in Britain in the 1920s. We therefore have an example of how the technical discourse of scientists and the public's understanding of science are linked. In this case, not only was the public's understanding of science informed by scientific discourse, public attitudes to science influenced the technical discourse of scientists.[4] We will argue that the electron and sub-atomic physics were presented as pure science, not just because this was the view held by physicists about the technical content, but also as a consequence of two other factors. The first was the need by a growing community of professional physicists to draw public attention to their work; the second was the development of a space for science in British public life.[5]

Physics—a New Discipline

In Britain relationships among physicists, government, and public altered as physics became an established academic discipline from the mid-nineteenth century. To encourage teaching and research in physics there were new institutions, such as the first university laboratories for research and teaching physics, and the Physical Society of London. Furthermore, a new journal, *Nature*, and the South Kensington Museum brought developments in physics to the notice of wider audiences.[6]

J. J. Thomson's own career illustrates the developments both in teaching and in research. He was a modern in that he was Professor of Experimental Physics and head of the Cavendish Laboratory, established by James Clerk Maxwell in Cambridge in the early 1870s. But Thomson was also Professor of Natural Philosophy at the older Royal Institution where his work followed the pattern established early in the nineteenth century by practitioners such as Faraday. There Thomson lectured to a largely lay audience who also acted as patrons of scientific research insofar as they paid for assistants to help Thomson with his research.[7] Later, when a wider audience of taxpayers came to provide the resources for scientific research, government agencies rather than individual researchers increasingly set the priorities. For example, the founding of the National Physical Laboratory in Britain around 1900 provided a new home for the important work of precision measurement. Consequently, physicists working in university physics laboratories could devote themselves to more speculative researches.[8]

For those caught up in the changes in physics around 1900, it was an exciting time. By comparison, physics research in the proceeding period seemed dull and pedestrian. In the introduction to his book on X rays first published in 1914 G. W. C. Kaye wrote

> 'In the early nineties, it was not infrequently maintained that the science of physics had put its house in complete order, and that any future advances would only be along the lines of precision measurement. Such pessimism has been utterly confounded by a sequence of discoveries since 1895 unparalleled in their fundamental nature and promise.'[9]

According to the historian Badash, only a small minority of physicists in the late 19th century held the view that physics was 'complete'. Nevertheless many of the next generation of physicists shared Kaye's vision.[10] Why should this happen? Why did some physicists in the early 20th century underestimate the ambitions of their predecessors? By adopting this opinion later workers would, in effect, magnify their own achievements. But apart from such considerations of personal ambition, they took this view of their predecessors because their own perceptions of their subject and themselves as physicists had changed.

In order to describe these changes, let us compare two exhibitions; the Exhibition of Pure Science at the British Empire Exhibition mentioned already and an earlier exhibition, the Special Loan Collection of Scientific Apparatus held at South Kensington, London, in 1876.

Exhibitions

The British Empire Exhibition held at Wembley in 1924–5 was one of a series of exhibitions in Britain inaugurated by the Great Exhibition of 1851. The emphasis in 1851 was on contemporary industry. Although modern scientific instruments were displayed they were presented as products of industry rather than being an illustration of current scientific research.[11] Soon afterwards, the Department of Science and Art and then the South Kensington Museum, were set up to promote the useful arts more systematically. However, the South Kensington Museum did not promote science and technology as effectively as some had hoped.[12] To encourage the Museum to develop in this direction, the Devonshire Commission stepped in. This Commission had set out in the 1870s to examine many aspects of scientific and technical education in Britain. Its chair, the Duke of Devonshire, was then Vice-Chancellor of the University of Cambridge and had personally funded the building the Cavendish Laboratory. Its secretary was Norman Lockyer, editor of the recently-founded journal, Nature.[13] In their fourth report the commissioners recommended the establishment of a collection of physical and mechanical instruments.[14]

Special Loan Exhibition of 1876

The Commission's proposal came to fruition in 1876 with the mounting of the Special Loan Exhibition of Scientific Apparatus at the South Kensington Museum. Its catalogue explained the purpose of the exhibition.

'Their Lordships stated their conviction that the development of the Educational, and certain other Departments of the South Kensington Museum, and their enlargement into a Museum somewhat of the nature of the Conservatoire des Arts et Métiers in Paris and other similar institutions on the Continent, would tend to the advancement of science, and be of great service to the industrial progress of this country.'[15]

However, in order to achieve this aim, the Special Loan Exhibition had to have a radically different structure from that of contemporary industrial exhibitions.[16] The Catalogue explained some of the differences:

'In International Exhibitions a certain amount of space is allotted to each country. These spaces are then divided by the commissioners of each country among its exhibitions, who display their objects ... as they consider the most advantageous .... the Exhibitions appeal naturally, more or less exclusively, to the industrial or trade-producing interests of those countries.

This was not the idea of the proposed Loan Collection at South Kensington. For that collection it was desired to obtain not only apparatus and objects from manufacturers, but also objects of historic interest from museums and private cabinets, where they are treasured as sacred relics, as well as apparatus in present use in the laboratories of professors. ... all objects ... were to be handed over absolutely to the custody of the Science and Art Department for exhibition; the arrangement being not by countries but strictly according to the general classification.'[17]

Thus the Loan Collection included historical material and arranged the objects in groups according to subject, not by country of origin. In due course eleven different countries provided some 4,600 exhibits, ranging from the famous Magdeburg hemispheres and air pump used by Otto von Guericke to over fifty different designs of barometer. These were arranged under twenty-one subject headings including Arithmetic, Geometry, Molecular Physics, Electricity, and Chemistry.[18] Historical instruments were used to illustrate the development of different types of scientific instruments. In effect these instruments were used to create a history for science, this history being seen primarily in terms of technical progress rather than physical theories.

The arrangement of the scientific instruments at the Special Loan Collection was by subject which helped to facilitate international comparisons. This was an explicit policy of the organisers. The catalogue explained their reasoning.

'It had been the intention from the first to give the Loan Collection an international character, so as to afford men of science and those interested in education an opportunity of seeing what was being done by other countries than their own in the production of apparatus, both for research and for instruction—an opportunity which it was hoped would be of advantage also to the makers of instruments.'[20]

Their plan was to display instruments for teaching, research, and industrial use from several countries; the audience was to be an increasingly significant group of specialists including teachers, researchers, and instrument makers.[21] By encouraging comparisons regarding government support of science the organisers hoped to influence the British government to increase its funding for science and to develop a science museum.[22] But these plans took some years to come to fruition.

To sum up, the 1876 Special Loan Collection of Scientific Apparatus displayed a broad range of scientific instruments and said little about

Figure 1. Magdeburg hemispheres from 1876 exhibition.[19]

scientific ideas and theories. It was firmly international in coverage and illustrated how science had developed over centuries as the result of international exchange. All these characteristics are in marked contrast to the later Exhibition of Pure Science at the British Empire Exhibition.

1924 British Empire Exhibition

By the late 19th century the usual pattern for international exhibitions was to concentrate on the industry of the nation holding the exhibition, and on the life and products of that nation's colonies.[23] The British Empire Exhibition of 1924 proved no exception, with a dazzling range of industrial and colonial exhibits. The Palaces of Engineering and Industry showed the products from a wide range of British firms. Dominions from India to Sierra Leone displayed their produce. The exhibition was not short of spectacle: historical naval battles were re-enacted, and the newly discovered Tutankamen's tomb was reproduced. The Ford Motor Company assembled cars and a full-scale coal mine was built under the site. The Board of Education showed a working illustration of the Sun trying, in vain, to set on the British Empire.[24] The Canadian Pavilion

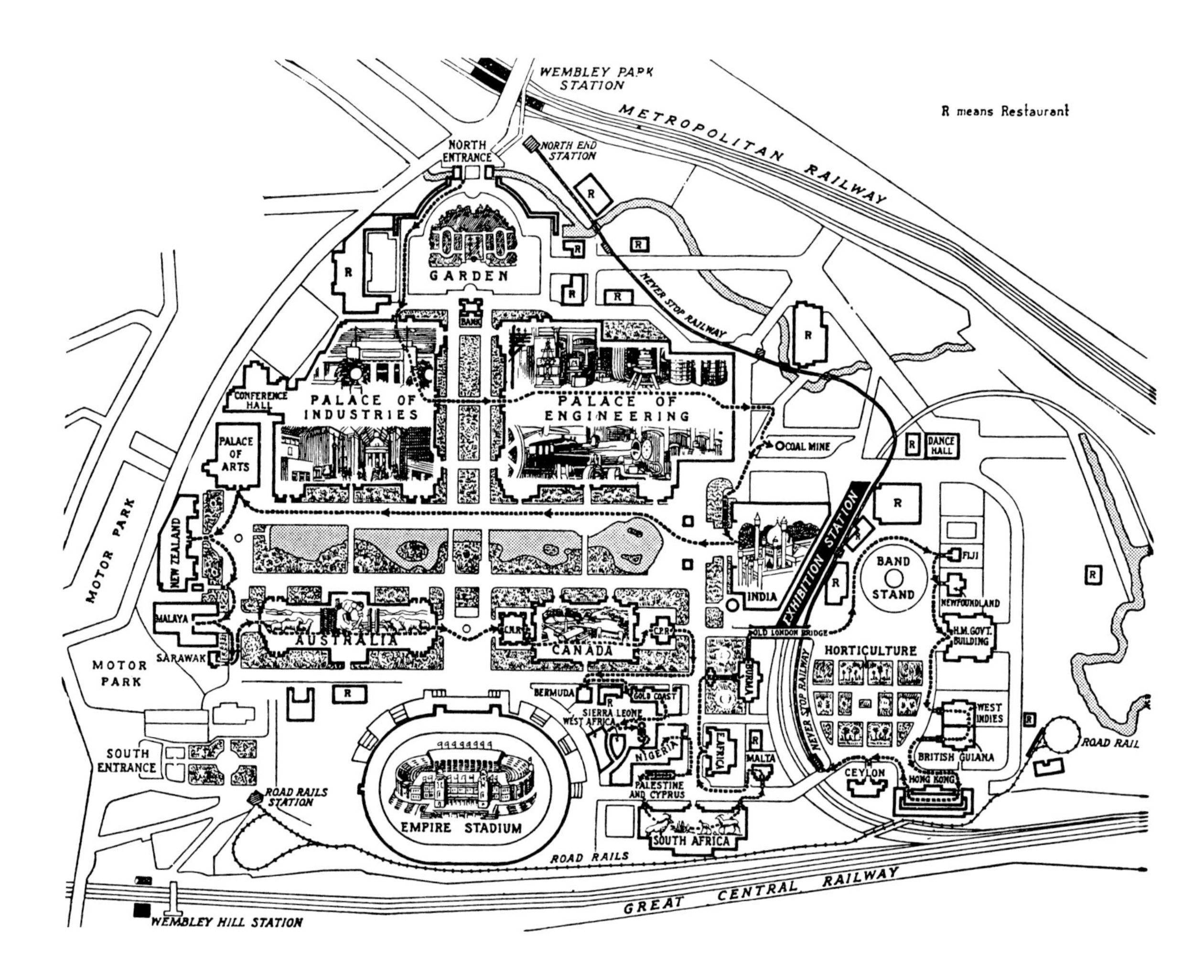

Figure 2. Map of the BEE.

featured a model of the Prince of Wales and his horse made out of butter in a special refrigerated case. Altogether, some 17.5 million visitors saw the exhibition during the six months of 1924 it was open, on average a staggering 100 thousand visitors a day.[25]

Many exhibits featured science and technology. The Palace of Engineering contained nine and a half acres of machinery from hundreds of engineering firms in a building 1000 feet long with 'no useless ornament'.[26] Collectively, the engineers were optimistic about their future prospects.

'For a hundred years there has not been such a need and such an opportunity as Engineering offers to the world today. A hundred years ago the condition of Europe was not unlike what it is at the present time. There had been a great and devastating war, in which practically all the civilised nations were engaged. Commerce had been brought to a stand-still, finance dislocated, and industry, which had hardly yet begun in the form in which we know it was languishing. Some new impetus was required to give a fresh start to the flow of industrial life. That impetus came from modern engineering, which took its rise about that time. And it came from Great Britain. This country was the cradle of modern engineering. The greatest of the early pioneers came from here. In that world crisis of a hundred years ago, it is not too much to say that England (and Europe with her) was saved by her engineers. They are ready to come to our help again to-day.'[27]

The beginning of modern engineering was coupled with recovery from a world crisis, with specific historical circumstances providing the conditions for a new profession to emerge.

As for the work of scientists, in the Palace of Engineering, the British Engineers' Association provided space for an exhibition by the National Physical Laboratory.

'... of special interest to scientists, is the exhibit and demonstrations by the National Physical Laboratory, who have been granted by the Association a free gift of stand space, at a cost of £1,400, in recognition of the valuable services rendered by the Laboratory to the engineering industry.'[28]

Nearby, the Palace of Industry housed displays of the companies of the chemical industry, which celebrated the work of chemists. A feature common to the exhibits mounted by chemists and engineers was the public acknowledgement of the importance of their work in wartime.

Exhibition of Pure Science

There was also a display of scientific instruments, the Exhibition of Pure Science that included Thomson's cathode ray tube. However, the circumstances here differed markedly from the nearby Palaces of Engineering and Industry. Furthermore, the way Thomson's tube was presented also differed from the way objects were presented in the Special Loan Collection almost fifty years before. The remainder of this paper will consider how this context, the Exhibition of Pure Science at the British Empire Exhibition, shaped the account of Thomson's tube.

This Exhibition of Pure Science was the first important public exhibition dealing with subatomic physics. Even as it was being viewed, the construction of a new building for the Science Museum (delayed by the First World War) was nearing completion. A display on science was planned for the new accommodation, which was officially opened in 1928.

A committee of the Royal Society had organised the exhibition, which was housed in the Government Pavilion and paid for by the Department of Overseas Trade. The chair of the Royal Society British Empire Exhibition committee was the prominent scientist, F. E. Smith. Its members included eminent scientists and electrical engineers, such as Oliver Lodge, W. H. Bragg, Richard Glazebrook (Director of the National Physical Laboratory from 1900–19), Arthur Schuster, Napier Shaw (Director of the Meteorological Office), Campbell Swinton and Henry Lyons (Director of the Science Museum).

Figure 3. The Pure Science Exhibition at the 1924 BEE. Courtesy of the Royal Society.

According to a draft for the guidebook,

'This exhibition, which has been arranged by the Royal Society at the request of H.M. Government, is intended to illustrate, so far as the space available will permit, recent British research in the fundamental branches of pure science.'[29]

As the exhibition was restricted to research currently carried out in Britain, only a selection of recent developments in science were covered. The core described the new understanding of the structure of the atom with many of the exhibits coming from the Cavendish Laboratory.[30] The draft continues

'By means of the instruments and apparatus used in the experiments, and by models, diagrams and photographs indicating the results, the work of Sir J J Thomson on cathode rays, leading to the discovery of the electron, of Sir Ernest Rutherford on the structure of the atom, of Sir William Bragg on crystal structure is illustrated; as well as recent work by many other distinguished scientists in the Physical, Engineering, Geophysical, Biological and other branches of science.'[31]

But preparing an exhibition on these aspects of physics posed considerable problems, for not only were individual electrons and atoms quite invisible, the details of atomic structure were unfamiliar to the public.[32] To overcome these hurdles, the exhibition organisers used a combination of original experimental equipment and working demonstrations.

'In many cases the exhibits are shown by the scientific men actually engaged in the work, and are supplemented by instrument(s) loaned by some of the leading firms of scientific instrument makers. The majority of the exhibits are working demonstrations. Benches, fitted with gas, water and electricity are provided, and a staff of scientific assistants are in attendance throughout the exhibition.'[33]

Although both the 1924 pure science exhibition and the Special Loan Collection of 1876 ostensibly dealt with science, there were important differences both in terms of subject matter and approach. The 1876 exhibition displayed static groups of similar instruments (fifty barometers) used for a wide range of purposes in teaching, research, and industry. The results and knowledge obtained by experiment were only of secondary concern. By contrast, in 1924, the Exhibition of Pure Science had different priorities. The organisers set out to explain the results of recent research about the structure of the atom. This knowledge came solely from experiments carried out in university research laboratories using specially developed and unique instruments. Some of these instruments were shown working in the 1924 exhibition to replicate actual experiments.[34]

Furthermore, in 1924, the arrangement of the individual exhibits was according to physical principles. The first exhibit was a chart showing the different electromagnetic waves that acted as a key to the arrangement of the principal section of the physical exhibits.[35] The next part of the exhibit concerned 'The Atom.' Starting from the electron, the exhibition displayed a new subject, subatomic physics, a subject it portrayed as 'pure science,' the outcome of work by professional researchers working in university laboratories and partially funded by the government.

Doubt and Uncertainty

Today the experiment Thomson carried out to measure the ratio of the charge, e, to the mass, m, of the electron, is regarded as the most significant part of his work on the electron. It is the aspect of Thomson's work most illustrated in textbooks.[36] This view was first given public exposure at the British Empire Exhibition.[37] It had the virtue of summarising complex judgements about an individual's contribution.[38] Thus the description used at the British Empire Exhibition gives credit to Thomson for his work at the Cavendish Laboratory in Cambridge (rather than the German physicists, Wiechert or Kauffmann, for example), thereby settling questions of personal, institutional, and national credit. It names the entity, the electron, provides it with an origin, thereby indirectly detailing properties such as its mass (much smaller than an atom) and electric charge (negative), and legitimates associated technical practices about how electrons are to be found (in cathode ray tubes containing gas at very low pressure).[39]

Such accounts also are useful within science, providing a way to cope with doubt and uncertainty and conveniently to date and to classify change and discovery. Physicists acting collectively settled debates about the reality of the particle, and once its existence was generally accepted, sometime after the development of the Rutherford-Bohr atom of 1911–13, a shorthand and retrospective account of the electron's origins becomes useful. Emphasising what is novel about the electron distinguishes it from the earlier theoretical speculations of Stoney, for example. In addition, the account provides plausible and definite answers when the details are either debatable or cannot be recovered.[40] The discovery account thus identifies a time, a place and particular people—providing the starting point for future commemorations.

Artefact as Icon

At the British Empire Exhibition, the electron, an invisible entity, and its history were made tangible through objects. As already mentioned J. J. Thomson exhibited his cathode ray tube described as the

> 'Apparatus by which the existence of electrons was detected and their mass and velocity measured'[41]

However, the history of this particular tube reveals how it became transformed into an icon. The diagram in Thomson's 1897 paper does not correspond exactly to this tube. Indeed, Thomson's paper makes clear

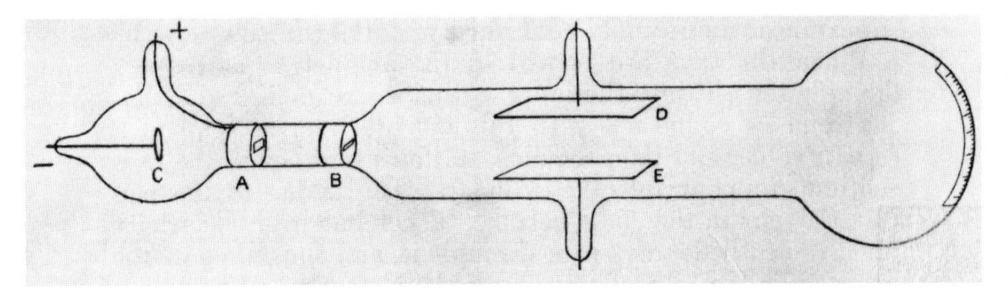

Figure 4. Diagram from J. J. Thomson's paper showing his experiment to measure the ratio of the charge to mass of cathode rays. Courtesy Trustees of the Science Museum.

that he made use of several different tubes in his experiments, in order to experiment with cathodes made from different materials, for example.[42]

Before this particular tube was shown at Wembley, it had been displayed at the Science Museum.[43] Information about how the tube was exhibited there comes from a letter written by the Secretary of the Royal Society British Empire Exhibition Committee in 1923.

'On the card attached to the apparatus in the Museum it is described as being that used by Sir J. J. Thomson for the measurement of the velocity of the Cathode [rays] and of the ratio m/e.'[44]

Significantly, this account describes the tube as an instrument for making measurements on cathode rays. So as a result of this tube becoming a museum exhibit, it came to stand for a range of experiments and apparatus. Furthermore, there is no explicit mention of the discovery of the electron. However, the measurements made with the tube could then, in a separate step, be interpreted as demonstrating the existence of the electron.

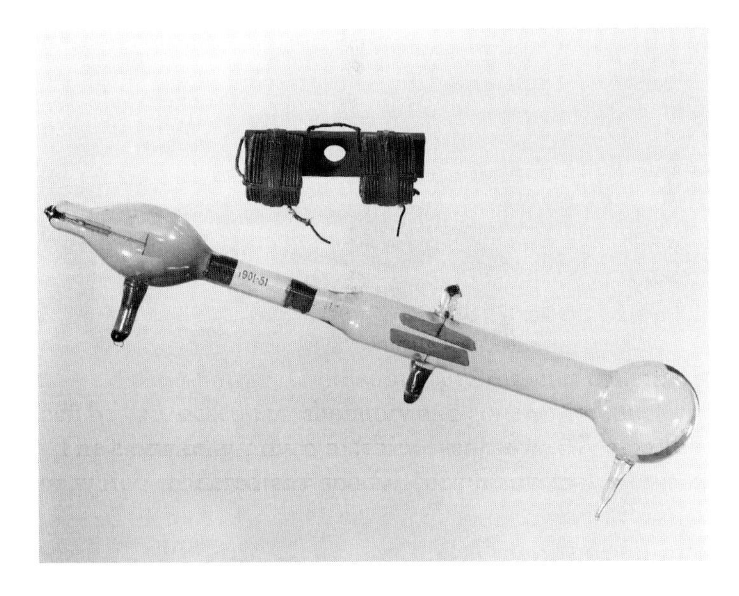

Figure 5. J. J. Thomson's tube from the Science Museum, Inventory Number 1901–51. Courtesy Trustees of the Science Museum.

Thomson himself was not immune to this revising of the historical account. When he arranged borrow this tube for the British Empire Exhibition, he wrote to Colonel Henry Lyons, the Director of the Science Museum,

'I have been asked to exhibit at Wembley the apparatus I used in my first experiments on the electron ... described I believe as apparatus for determining the velocity and the value of e/m for cathode rays.'[45]

So in 1924 Thomson viewed his earlier experiments as involving electrons (rather than 'corpuscles' as he called them for many years). Furthermore, he now stated the ratio as e/m rather than in the inverse form, m/e, he had used in his original paper and which had then been quoted in the label at the Science Museum.

By the time of the British Empire Exhibition in 1924, however, the account of Thomson's work had been pared down and simplified until the device becomes the tube used for the experiment. The artefact was now the actual device used for detecting the existence of an invisible entity, the electron.

Following this account of the electron at the Exhibition of Pure Science, the next section dealt with

'Some Aspects of Radio-Activity and their Bearing on Atomic Structure'.

was provided by Sir Ernest Rutherford, Thomson's successor as head of the Cavendish. This section included material on radium emanation (radon), the nature of the alpha particle, the scattering of alpha-particles, disintegration of elements, beta and gamma rays, and the work of Moseley on Atomic Number. After that came exhibits by D. R. Hartree, 'Models Illustrating Atomic Structure' (which provided a snapshot of atomic structure according to 'Old Quantum Mechanics' and immediately before the development of 'New Quantum Mechanics'), and a model of the structure of rocksalt from W. L. Bragg. Also illustrated were C. T. R. Wilson's cloud chamber method and Aston's Mass Spectrograph.[46]

Not only are contemporary scientific theories about the structure of the atom used to order the exhibits, historical events are fitted to this pattern too. The electron is the subject of the first exhibit, and its discovery is the first step in the historical story that unfolds, a story of discovery and progress in both knowledge and technique.

Empire

The Exhibition of Pure Science illustrated the research carried out in Britain. However, some of the scientists who contributed to it, including W. H. Bragg and Ernest Rutherford, had links with the British Empire which was apposite, given the exhibition was part of the British Empire

Exhibition. In Rutherford's case, not only had he been born in New Zealand, he had been professor of physics in McGill University, Montreal, before leaving for Manchester and later Cambridge.[47] Indeed Thomson's students, such as Rutherford, occupied many of the chairs of physics in the British Empire.[48]

But there were other links between science and Empire, which the pure science exhibition illustrated. The exhibition was shown in 1924, in the aftermath of the First World War, and, as a consequence, references to nation and empire replaced the explicit internationalism of the earlier Special Loan Exhibition.[49] As was pointed out above, the account given of the electron gave credit to Thomson but not to his German colleagues such as Wiechert or Kauffmann. Similarly, the part of the Exhibition of Pure Science dealing with X-rays made no mention of the German physicist, Wilhelm Röntgen, their discoverer, but instead mentioned Charles Barkla, an English physicist who discovered the characteristic X-rays emitted by different elements. Clearly these accounts owed much to contemporary political circumstances.

Although contact with scientists in Germany was restricted, they did not cease completely. When the time came to return the exhibits from the British Empire Exhibition to the Cavendish Laboratory, responsibility for the arrangements lay with Patrick Blackett. However, a disgruntled James Chadwick had to step in for Blackett who had gone to Germany in the meantime to work in James Franck's laboratory in Göttingen.[50]

The Professional Identity of Physicists

Another feature of the pure science exhibition at the British Empire Exhibition was the strong emphasis on the idea of research as worth doing for its own sake. There were no mention of any practical applications with the single exception of wireless waves. This work was carried by professional university researcher in physics, individuals doing research for its intellectual worth, and not engaged primarily in teaching (the traditional university occupation) nor in solving more mundane practical problems (the physicist in industry).

Concern among physicists about their professional identity became widespread in the early 1920s. The first society for professional physicists in Britain, the Physical Society of London, had been formed in February 1874 (significantly during the planning stages of the Special Loan Collection). Subsequently, the growth of university physics departments and the setting up of the National Physical Laboratory had greatly increased the number of professional physicists.[51] However, it was the experience of the First World War that convinced many physicists of the need for a professional body to represent their interests. For they felt that they had fallen behind their colleagues, the chemists, in terms of professional status.

Furthermore, the setting up the Department of Scientific and Industrial Research to channel Government money into scientific research offered new opportunities for physicists.[52] The inaugural meeting of the Institute of Physics was held in April 1921. This brought together representatives of the Faraday Society, the Optical Society, and the Physical Society of London. Sir Richard Glazebrook, recently retired as Director of the National Physical Laboratory, was its first President.[53]

At this Inaugural meeting, the first Honorary Fellow, J. J.Thomson drew attention to the increase in the numbers of physicists in Britain. In 1870, when he began his own career, he thought there had been no more than 100 whereas by 1920 he estimated their numbers to have grown to between 800 and 1000, a total which included science teachers with an interest in research. These were the potential members of the new Institute, and within the first year 300 had joined.[54] Now physicists in Britain had an organisation to pursue their professional interests.

Thus there was a significant conjunction in 1924. Prominent members of the new Institute of Physics, concerned about promoting the professional interests of physicists, arranged part of the Exhibition of Pure Science. They had an excellent opportunity to bring the work of their members to the attention of the public.

Pure Science

Some characteristics of a distinctive public identity for physicists emerge at the 1924 exhibition. Although many worked in industry, what was newly displayed in 1924 was pure science; a term that applied to work carried out by scientists in universities where they were free to research. A strong advocate of the value of curiosity-driven research was J. J. Thomson. In 1916, Thomson, then President of the Royal Society, explained his views at a meeting where the Government announced the establishment of the Million Fund to aid co-operative research in industry.[55]

'By research in pure science I mean research made without any idea of application to industrial matters but solely with the view of extending our knowledge of the Laws of Nature. I will give just one example of the "utility" of this kind of research, one that has been brought into great prominence by the War—I mean the use of X-rays in surgery. Now, not to speak of what is beyond money value, the saving of pain, or, it may be, of life to the wounded, and of bitter grief to those who loved them, the benefit which the State has derived from the restoration of so many to life and limb, able to render services which would otherwise have been lost, is almost incalculable. Now, how was this method discovered? It was not the result of a research in applied science starting to find an improved method of locating bullet wounds. This might have led to improved probes, but we cannot imagine it leading to the discovery of the X-rays. No, this method is due to an investigation in pure science, made with the object of discovering what is the nature of Electricity.'[56]

Thus, for Thomson, research in pure science carried out for curiosity's sake, may well be worth doing for its practical benefits—it is just that they are impossible to foresee. Thomson's argument was twofold; the researchers had the moral advantage of a 'pure' intention but their research might also provide practical advantages for humanity. Clearly, this vision of research in pure science fitted the work of researchers in universities who are 'free' to follow this kind of research, and the prospect of funding from the Department of Scientific and Industrial Research.[57] The notion of pure research distinguished what scientists in universities did from the types of research done by researchers working for particular ends, on standards or under contract at the NPL, or by electrical engineers, for example.[58]

There is another feature of the work of the 'pure' scientists is apparent at the British Empire Exhibition. There the work of Thomson, Rutherford, Bragg, and others, was represented as pure science and there was no mention of their wartime research for the Admiralty and other agencies. In contrast, elsewhere in the British Empire Exhibition, the chemists, engineers, and the physicists from the National Physical Laboratory, were forward in bringing their contribution to the war effort to public notice. But if applied researchers draw attention to the results of their labours in wartime, one function of the idea of pure science is to bring forward an idea of science as knowledge without destructive consequences. Thus pure science is not just about the lack of applications; it is about purity stemming from a lack of destructiveness.[59] The great irony, of course, is the subatomic physics displayed as Pure Science at the British Empire Exhibition developed into nuclear physics that had a great bearing on warfare during World War II with the development of nuclear weapons.

This irony was particularly acute for the physicists at the Cavendish. For they were strongly identified with the idea of pure science as knowledge for knowledge's sake, a factor which must have encouraged them to take part in the Exhibition of Pure Science.[60] A further example of the irony that the 'pure science' of atomic physics developed into something destructive comes from an interview with James Chadwick shortly after he published the results of his first experiments on the neutron at the Cavendish early in 1932.

'Dr Chadwick described his experiments as the normal and logical conclusion of the investigations of Lord Rutherford 10 years ago. Positive results in the search for 'neutrons' would add considerably to the existing knowledge on the subject of the construction of matter, and as such would be of the greatest interest to science, but, to humanity in general the ultimate success or ortherwise of the experiments that were being carried out in this direction would make no difference.'[61]

But if physicists had reasons for portraying atomic physics as pure science, then the public was receptive to this approach for related reasons. While questions about the structure of matter have fascinated both researchers and public alike through the centuries, around 1920, it seems the public took a greater interest in the structure of the atom for then it was a desperately needed escape from the problems created by the war.[62]

Conclusion

By examining how a tube used by J. J. Thomson was displayed to the public in 1924, we see how the image of physics held by both physicists and members of the public had changed since 1800. As research in physics became an activity funded by Government using money from taxes, physicists developed a public identity for their discipline. While the changes were due in large part to the new ideas and discoveries made by physicists around that time, a factor emphasised in accounts by physicists, the argument presented in this paper is that the changes were also the result of a new institutional structure. In the process, Thomson's tube and the electron became emblems of pure science.

What have been emphasised here are some of the events that led to the establishment of a permanent public space for science, what became the Science Museum in London. While this public space displayed images of science and technology that were the outcome of private discussions in government circles, professional bodies of scientists and engineers, and trade associations what was new for the public was a vision of pure science.

Notes

1. See D. Halliday and R. Resnick, *Physics* (New York, 1966), Part II, p. 835. Thomson published the results of this experiment in October. J. J. Thomson, 'Cathode Rays,' *Philosophical Magazine* 44 (1897), 293–316. The previous April Thomson had first made public his idea that subatomic particles existed at a lecture in the Royal Institution. These 'corpuscles,' as he called them, had negative electric charge and were constituents of all matter.
2. See, for example, S. Weinberg, 'The First Elementary Particle,' *Nature* 386 (20 March 1997), 213–215. T. Arabatzis, 'Rethinking the "Discovery" of the Electron,' *Stud. Hist. Phil. Mod. Phys.* 27(4) (1996), 405–435.
3. Royal Society, *Phases of Modern Science* (London, 1926), p. 146.
4. See P. Forman, 'Weimar Culture, Causality, and Quantum Theory 1918–1927: Adaption by German Physicists and Mathematicians to a Hostile Intellectual Environment.' *Historical Studies in the Physical Sciences* 3 (1971), 1–115.
5. We can now understand why commentators who have only considered the historical evidence from around 1897 have found this account of the electron wanting, for the conditions of the 1920s also have to be considered. See, for example, T. M. Brown, A. T. Dronsfield, and J. S. Parker, 'Did Thomson Really Discover the Electron?' *Education in Chemistry* (May 1997). Others have made less extravagant cases. Pais records in meticulous detail the several experimenters who have claims to have found the electron. His account offers a traditional conclusion with a new twist; Thomson did discover the electron, not in 1897, but in 1899 when he measured the charge on the electron, thus eliminating, in Pais's view the last significant obstacle to identification. A. Pais, *Inward Bound: Of Matter and Forces in the Physical World* (Oxford, 1986).
6. See D. S. L. Cardwell, *The Organisation of Science in England* (London, 1972). R. Sviedrys, 'The Rise of Physics Laboratories in Britain,' *Historical Studies in Physical Sciences* 7 (1976), 405–436. I. R. Morus, 'Manufacturing Nature: Science, Technology and Victorian Consumer Culture.' *BJSHS* 29 (1996), 403–434. G. Gooday, 'Precision Measurement and the Genesis of Physics Teaching Laboratories in Victorian Britain.' *BJHS* 23 (1990), 25–51. P. Forman, J. L. Heilbron, et al., 'Physics circa 1900: Personnel, Funding and Productivity of the Academic Establishments.' *Historical Studies in Physical Sciences* 5 (1975).
7. Thomson was Professor of Natural Philosophy there from 1905. His assistants were G. W. C. Kaye and then Francis Aston who won the Nobel prize for his work on isotopes.
8. See M. J. Nye, *Molecular Reality: A Perspective on the Scientific Work of Jean Perrin* (London, 1972).

9. G. W. C. Kaye, *X Rays* (London, 1923), p. xix.
10. L. Badash, 'The Completeness of Ninteenth-Century Science.' *Isis* 63 (1972), 49–58. T. S. Kuhn, *Black-Body Theory and the Quantum Discontinuity*, 1894–1912 (Oxford, 1978). See also S. Schaffer, 'Utopia Limited: On the End of Science.' *Strategies*(4/5) (1991), 151–181.
11. C. H. Gibbs-Smith, *The Great Exhibition of 1851* (London, 1981). J. A. Bennett, *Science at the Great Exhibition* (Cambridge, 1983). J. A. Bennett, *The Divided Circle* (Oxford, 1987). D. Follett, *The Rise of the Science Museum under Henry Lyons* (London, 1978). R. Brain, *Going to the Fair. Readings in the Culture of Nineteenth-Century Exhibitions* (Cambridge, 1993).
12. See S. Forgan and G. Gooday, 'Constructing South Kensington: The Buildings and Politics of T. H. Huxley's Working Environments,' *BJHS* 29 (1996), 435–468.
13. A. J. Meadows, *Science and Controversy: A Biography of Sir Norman Lockyer*, (London, 1972). D. S. L. Cardwell, *The Organization of Science...*, pp. 119–126. *A History of the Cavendish Laboratory 1871–1910* (London, 1910).
14. See *Royal Commission on Scientific Instruction and the Advancement of Science, Fourth Report* (London, HMSO, 1874), p. 23, recommendation IX.
15. *Catalogue of the Special Loan Collection of Scientific Apparatus at the South Kensington Museum* (London, 1876). p. vii.
16. *Catalogue of the Special Loan Collection of Scientific Apparatus at the South Kensington Museum*, (London, 1876).
17. Ibid. *Catalogue of the Special Loan Collection*, 2 vols, 2nd edition (London, 1876), vol 1 p. ix.
18. The countries providing exhibits were the United Kingdom, Austro-Hungarian Empire, France, Germany, Holland, Italy, Norway, Russia, Spain, and Switzerland. That year the United States held the Centennial Exhibition in Philadelphia. At the time the preoccupation of the organisers of the Special Loan Collection with scientific instruments was shared by others. The new physics laboratories then being constructed at universities had to be equipped and James Clerk Maxwell, for one, took great trouble to acquire suitable instruments for the Cavendish Laboratory. Similarly, the physicists who in 1874 founded the Physical Society of London, the first specialist society in Britain for physicists, made a point of having novel demonstration experiments at their meetings. See G. Gooday, 'Precision measurement and the genesis of physics teaching laboratories in Victorian Britain,' *BJHS* (1990), 23: 25–51.
19. These are in the collections of the Deutsches Museum. While they were in South Kensington, copies were made.
20. *Catalogue of the special loan collection of scientific apparatus at the South Kensington Museum* (London, 1876), p. viii. To encourage research, there were conferences held while the Collection was on display. At one, James Clerk Maxwell gave a paper on scientific instruments.
21. See M. E. W. Williams, *The Precision Makers: A History of the Instruments Industry in Britain and France, 1870–1939* (London, 1994).
22. F. M. Turner, 'Public Science in Britain, 1880–1919,' *Isis* 71(1980), 589–608.
23. R. Brain, *Going to the Fair. Readings in the Culture of Nineteenth-Century Exhibitions* (Cambridge, 1993). Tony Bennett, *The Birth of the Museum: History, Theory, Politics* (London, 1995). P. Greenhalgh, *Emphemeral Vistas: A History of the Expositions Universelles, Great Exhibitions and World's Fairs, 1851–1939*, (Manchester, 1988).
24. This was written as Hong Kong was handed over to China.
25. The earlier Paris Exhibition had even more visitors, 44 million.
26. *Engineering Sectional Catalogue British Empire Exhibition* (London, 1924), pp. 14–5.
27. Ibid, p. 13.
28. Ibid., p. 48. The NPL also provided some exhibits for the exhibition on pure science.
29. Royal Society archives, British Empire Exhibition Committee, Box 7 Department of Overseas Trade, File B, Draft for guide book, dated 22 January 1924.
30. The physics exhibits dealing with the atom and radiations from Gamma Rays to Wireless Waves takes up more space in the catalogue than the other sciences, namely Geophysics, Zoology, Botany and Physiology.
31. Royal Society archives, British Empire Exhibition Committee, Box 7 Department of Overseas Trade, File B, Draft for guide book, dated 22 January 1924.
32. Recalling a striking remark made by Sir J. J. Thomson, Oliver Lodge wrote '...though the present century had been extraordinarily prolific in discoveries in Natural Science, especially in the more physical branch thereof, there was very little popular knowledge or understanding thereof.' O. Lodge, *Atoms and Rays*, (New York, 1924), p. v.

33. Royal Society, *Phases of Modern Science* (London, 1925?), p. 34. Of course, working demonstrations had long been used for teaching.
35. Ibid, pp. 161–2.
36. D. Halliday and R. Resnick, *Physics* (note 1).
37. When the Post Office in Britain declined to issue special stamps to mark the electron centenary, preferring instead to issue stamps honouring Enid Blyton, the children's author, some members of the Institute of Physics were incensed. R. Cooter and S. Pumfrey, 'Separate Spheres and Public Places: Reflections on the History of Science Popularization and Science in Popular Culture,' *History of Science* 32 (1994), 237–267. '1947 Jubilee Exhibition: Life, the Universe and the Electron,' Science Museum (1997).
38. See R. A. Millikan, *The Electron* (Chicago, 1963). Millikan's case also illustrates some personal and national characteristics by comparison with Thomson's. Millikan regarded his oil drop apparatus as so much junk once he had finished with it whereas Thomson's tubes have gone to museums. Millikan entered into an arrangement to be buried at Forest Lawn as one of their 'Immortals.' Thomson was buried in Westminster Abbey.
39. See R. B. Leighton, *Principles of Modern Physics*, (New York, 1959), p. 624. See also A. Pais, *Subtle is the Lord: The Science and Life of Albert Einstein*, (Oxford, 1982); T. Arabatzis, 'Rethinking the "Discovery" of the Electron.' *Stud. Hist. Phil. Mod. Phys.* 27(4) (1996), 405–435.
40. Stoney coined the term electron, see J. O'Hara, *George Johnstone Stoney and the Conceptual Discovery of the Electron*, (Dublin, 1993). S. Schaffer, 'Scientific Discoveries and the End of Natural Philosophy,' *Social Studies of Science* 16 (1986), 387–420. W. C. Dampier, *The Recent Development of Physical Science* (London, 1904). E. Hobsbawm, and T. Ranger (eds.) *The Invention of Tradition*, (Cambridge, 1983).
41. Royal Society, *Phases of Modern Science* (London, 1925), p. 146. Thomson wrote this caption himself. See J. J. Thomson to Martin (?), Secretary of the Royal Society British Empire Committee, April 4 1924, Box 3, British Empire Exhibition Committee, Royal Society Archives. In addition Thomson provided a modified Perrin tube he had also used in his experiments. This Thomson had used to repeat an experiment originally performed by Perrin which served to identify that cathode rays carried a negative electric charge. The Perrin tube is in the museum of the Cavendish Laboratory.
42. Apart from the diagram there are few details in the paper which would serve to identify the tube exactly. Thomson's notebooks detailing his experiments in this period are not present in the collections of his papers. Thomson also introduced different gases into his tubes. The idea being that if neither the material of the cathode nor the trace of gas left in the tube made any difference to his measurements of e/m, then his corpuscle was more likely to be a constituent of all matter. Today there are other similar tubes in the Cavendish Laboratory and one in the Deutsches Museum that belonged to Thomson. See I. J. Falconer, *Apparatus from the Cavendish Museum* (Cambridge, 1980).
43. In fact, originally the tube had gone from the Cavendish Laboratory to what was then the Victoria & Albert Museum, Science Collections in 1901. The Museum's annual report for that year described the acquisition as 'Apparatus for measuring the velocity of the cathode rays.' Only in 1909 did the V&A and the Science Museum become separate organisations. J. Physick, *The Victoria and Albert Museum: The History of its Building* (London, 1982), p. 24.
44. The Secretary of the Royal Society's British Empire Exhibition committee to P. M. S. Blackett, 22 December 1923. Royal Society British Empire Exhibition 1924 Box 3 Cambridge file. The word rays has been added in ink to the typescript. Blackett, acting for Rutherford, was responsible for the Cavendish Laboratory's contribution to the exhibition.
45. See File ScM 177.
46. The Science Museum even today has a number of these items on display, a direct legacy of the British Empire Exhibition.
47. He was also one of the first holders of a scholarship established by the Commissioners of the 1851 Exhibition, a scholarship which enabled him to study in Cambridge. A further reason for linking science and empire was that in the 1920s the Department of Scientific and Industrial Research in Britain and its equivalent in Australia, the CSIR, began to coordinate their programmes for research. See C. B. Schedvin, *Shaping Science and Industry: A History of Australia's Council for Scientific and Industrial Research, 1926–49* (Sydney, 1987). J. Morrell and A. Thackray, *Gentlemen of Science: Early Years of the British Association for the Advancement of Science* (Oxford, 1981). The BAAS held meetings in different parts of the empire.

48. 'During the whole of his tenure of the Cavendish chair, Thomson was the active head and inspirer of a great research school. The institution of the status of "advanced student" at Cambridge in 1895 led to a great increase in the numbers of young graduates of other universities who came to work in his laboratory; with the consequence that in a few years nearly all the important chairs of physics in the British Empire were filled by his disciples.' E. T. Whittaker, A *History of the Theories of Aether and Electricity: The Classical Theories* (New York, 1951), p. 366.
49. See A. G. Cock, 'Chauvinism and Internationalism in Science. The International Research Council, 1919–26,' *Notes and Records of the Royal Society* 37 (1983), 249–288. B. Schroeder-Gudehus, 'Nationalism and Internationalism,' in R. C. Olby, G. N. Cantor, et al. (eds.), *Companion to the History of Modern Science* (London, 1990), pp. 909–19.
50. See James Chadwick to Martin, secretary of the Royal Society Committee, 18 November 1924, in Royal Society Archives, Cambridge, British Empire Exhibition 1924, Box 3..
51. See E. Pyatt, *The National Physical Laboratory: A History* (Bristol, 1983).
52. See D. S. L. Cardwell, *The Organization of Science ....* (note 6). I. Varcoe, *Organizing for Science in Britain* (Oxford, 1974).
53. M. G. Ebison, *The Development of the Physics Profession Linked to the Education and Training of Physicists to 1939* (Salford, 1991), p. 334. The new Institute was not a trade union. cf Association of Scientific Workers.
54. Ibid., p. 341. The new Institute soon collaborated with the NPL to produce the *Journal of Scientific Instruments.*
55. See D. S. L.Cardwell, *The Organizaton of Science....* (note 6).
56. L. Rayleigh. *The Life of Sir J. J. Thomson O. M.* (Cambridge, 1942), pp. 198–89.
57. See K. Gottfried and K. G. Wilson, 'Science as a Cultural Construct.' *Nature* 386 (10 April 1997), 545–47, to see an example of how the idea of pure science is still used today.
58. R. Appleyard, *The History of the Institution of Electrical Engineers 1871–1931*(London, 1939).
59. See J. A. Hughes, *The Radioactivists: Community, Controversy and the Rise of Nuclear Physics.* PhD dissertation, Department of History and Philosophy of Science, Cambridge, 1993.
60. Interestingly, some of the growth of physics in Cambridge was the direct outcome of the 1851 exhibition, for a portion of the proceeds of the exhibition was used to fund scholarships. Of the scholarships awarded between 1891–1895, 29 were held by physicists, of whom only two went to Cambridge. However, between 1896 and 1921, 103 physicists had scholarships, of whom 60 went to Cambridge. L. Rayleigh, *The Life of Sir J. J. Thomson.*
61. Interview with James Chadwick reported in *The Times* (Monday, 29 February 1932). A later example comes from an exhibition, 'Atom Tracks,' at the Science Museum in 1936–37 which showed the use of the cloud chamber developed by C. T. R. Wilson.
62. For similar reasons, Dirac argued the public showed great interest in relativity in that period. R. Williamson (ed.), *The Making of Physicists* (Bristol, 1987), p. 6.

Hartmut Petzold

Wilhelm Cauer and his Mathematical Device

In the latter half of the 1920s Wilhelm Cauer (1900–1945) developed his plan for an electromechanical machine to solve linear equations. By the end of the next decade he would be acquainted with four similar projects which had emerged autonomously in different parts of the world. These other machines differed from each other and from his in both concept and technological realization, but they all were capable of calculating complicated formulas in one step or in an automatically performed sequence of steps. Together they can be seen as representative elements of the pre-history of the stored-program computer. They confirm the existence of a growing demand for calculating machines which could exceed the capabilities of contemporary desk top calculators to add and multiply. They indicate that some common aspects of the scientific-technological world produced identical questions and problems on the European continent, in Great Britain and on the other side of the Atlantic, in spite of considerable differences in cultural and political situations.

In the epilogue to his excellent analysis of the situation prior to the modern electronic digital computer William Aspray considers the mosaic of what are accepted as the most important calculating instruments and machines. He then states that 'the rich connections between the technology we have set out in this book and the electronic, stored-program computer' can only be suggested, and that the 'incomplete understanding of these connections' even 'may seem odd to some readers.'[1] In that same book, Allan G. Bromley's chapter on 'Analog computing devices' presents the accepted background for the story being told here. He concludes that the listed machines 'were all of an ad-hoc nature and did not lead to any general synthesis or the emergence of a general class of machines,'[2] with the exception of Vannevar Bush's Differential Analyzer and the mechanical Gunnery Computers. Bromley doesn't mention Cauer's machine, which is appropriate since, in his view, it would be another of the broad spectrum of 'ad hoc' projects one can find scattered through contemporary journals and texts.[3]

In this essay I take another path. Cauer was not a member of some prominent community of physicists and did not achieve the kind of professional success that has drawn the attention of historians. The written sources are therefore weak, and I have to tell the story of his machine using other material. I employ a method in which I centerpiece Cauer against accounts of four comparable machines with which he was familiar. I can describe an informal connection even though I cannot verify exactly Cauer's opinions. Thus I might be able to say that he was impressed by a particular machine without knowing whether it was favorably or unfavorably. And I can infer that this knowledge of the work of others influenced him as he moved from an initial stage where he considered only mathematical, physical, and technical problems, to a second stage where he was concerned with comfortable, reliable, and semiautomatic handling, to a final stage where he appeared to worry about something like market analysis. By which time he had seen the big machine projects in the United States. So this essay can be seen as a reconstruction of Wilhelm Cauer's subjective attitude. And, although the evidence is limited, I am able to argue that his work was not completely 'ad hoc' but that it was at least partially linked to the work of others.

At the same time, I describe how problem-solving by means of mathematical models, formulae, and algorithms grew in the 1920s to become a critical difficulty, especially for programs in science and engineering. Many of the contemporary problems culminated in mathematical systems of linear equations which had to be calculated by means of the Gauß algorithm. This was generally true for science and engineering in western societies in the first half of the 20th century. Wilhelm Cauer's particular contribution was the suggestion of an electro-mechanical solution of this algorithm.

In addition, this essay is an attempt to interpret technological artefacts as well as written documents and diagrams as historical testimony. It should offer a possible historical interpretation of an artefact, by enlarging the focus to an ensemble of comparable artefacts. Unfortunately, not all machines described still exist. Those by Hull and IBM (Columbia) are preserved at the National Museum of American History. Details of the others come from written and printed material of differing quality.[4]

Wilhelm Cauer's Machine Project

Wilhelm Cauer was born in 1900 as the youngest child of a well-situated academic Berlin family. His father was the first professor for railway practice at the Technische Hochschule (TH) Charlottenburg. Many other members of the Cauer family were prominent scholars. Three of his five sisters had doctoral degrees. Wilhelm Cauer's son has termed his ancestors typical 'Bildungsbürgers,'[5] an element of society that played an important role in the German Kaiserreich.

Cauer was educated in electrical engineering, physics and mathematics at the Technische Hochschule Charlottenburg, and at the universities of Berlin and Bonn. He graduated from the TH with a diploma in Technische Physik and then enrolled at the laboratory of the Berlin telecommunication company Mix und Genest. In 1925 he returned to the TH as an assistant to Georg Hamel at the Institute of Mathematics and Mechanics.[6] This institute was the theoretical and mathematical center for all engineering disciplines at the TH.[7] At Berlin and elsewhere, mechanical statics and dynamics, particularly gas dynamics, represented the most advanced theoretical engineering discipline of the time. Still, until after the Second World War, Hamel's textbooks for mathematical mechanics played an important role in the education and working style of German engineers.[8] So, Hamel's institute can be seen as a good place for an open-minded young assistent to learn how to handle one of the great problems of mathematical engineering: how can engineers get concrete numbers from the enormous analytical formulae which were offered by mathematics and theoretical physics?

At the end of 1925 Cauer completed his doctoral dissertation 'Die Verwirklichung von Wechselstromwiderständen vorgeschriebener Frequenzabhängigkeit'. This was the first step into a field, where he would exercise great influence to the engineering of the 20th century, and where he would work until his death.[9] At Mix und Genest he had learned how to deal with patents. Cauer registered a considerable number of patents during his short life. Indeed, it seems that his first approach to the development of mathematical apparatus was to seek a patent,[10] even if he had no plan to build it.

It seems that by 1927 he already had an idea of how to solve systems of linear equations by means of an electrical device, reproducing the Gauß algorithm to transform the matrix in its triangular form. Subsequently it would be simple to calculate the unknowns with a common mechanical desk calculator.[11] Presumably he tried to obtain a patent, but there may have been problems with the application. It is not mentioned in the schedule of Cauer's patents, which was compiled after his death.[12] There is in fact only one published description of the calculating device. Cauer wrote it at the end of 1934, at the conclusion of the whole project.[13]

The crucial concept imbedded in his device represented an alternative to existing digital mechanical calculating machines and also to slide rules and cylinders. The digital numerical solution of the algorithm required a long series of calculations where the results from one were the factors for the next, and so on. So the error inherent in rounding numbers, and in multiplying and dividing rounded numbers, grew with the number of the equations. Cauer's concept incorporated an analogue device, the precision of which, he believed would be sufficient for most engineering problems. The critical element was an electrical Wheatstone bridge with variable decimal resistors. The bridge circuit represented the analog part, and the decimal resistors the digital part of this hybrid device.[14]

In Cauer's apparatus the coefficients were represented by resistors. The calculation proceeded in a step-by-step elimination of the equation coefficients by balancing the different bridge circuits. So it was not necessary to measure the currents absolutely. The system of linear equations

$$a_{11 \times 1} + a_{12 \times 2} + a_{13 \times 3} + a_{14} = 0$$
$$a_{21 \times 1} + a_{22 \times 2} + a_{23 \times 3} + a_{24} = 0$$
$$a_{31 \times 1} + a_{32 \times 2} + a_{33 \times 3} + a_{34} = 0$$

could be transformed into the triangular system of these equations.

$$a_{11 \times 1} + a_{12 \times 2} + a_{13 \times 3} + a_{14} = 0$$
$$a'_{22 \times 2} + a'_{23 \times 3} + a'_{24} = 0$$
$$a'_{33 \times 3} + a'_{34} = 0$$

First the resistors a_{11}, a_{12}, a_{13}, a_{14} were set as the coefficients of the first equation. The switches U changed the sign. In the diagram parallel connections symbolize the positive sign, crossing connections the negative ones. Now the upper resistor row is set with the coefficients of the second equation. The current I_1 through the circuit is tuned by changing R_1 so that the voltage sum $a_{21}I_0 + a_{11}I_1$ becomes 0. This state occurs if the galvanometer G indicates no current. Then in the next row of resistors a'_{22}, a'_{23}, a'_{24} is set so that the current is constant, and a'_{22} is set by changing R_2 so that the voltage sum $a_{22}I_0 + a_{12}I_1 1 + a'_{22}I_2$ is 0. The galvanometer G again indicate no current when S is set to 2. And so on.

Cauer stated that he had calculated several complete systems using this method at Professor Max Reich's Institut für angewandte Elektrizität at the University of Göttingen.

Cauer began constructing an automatic version of his machine for three equations with three unknowns during the first half of 1930, but he could not complete it because of lack of funds. We know Cauer's circuit design

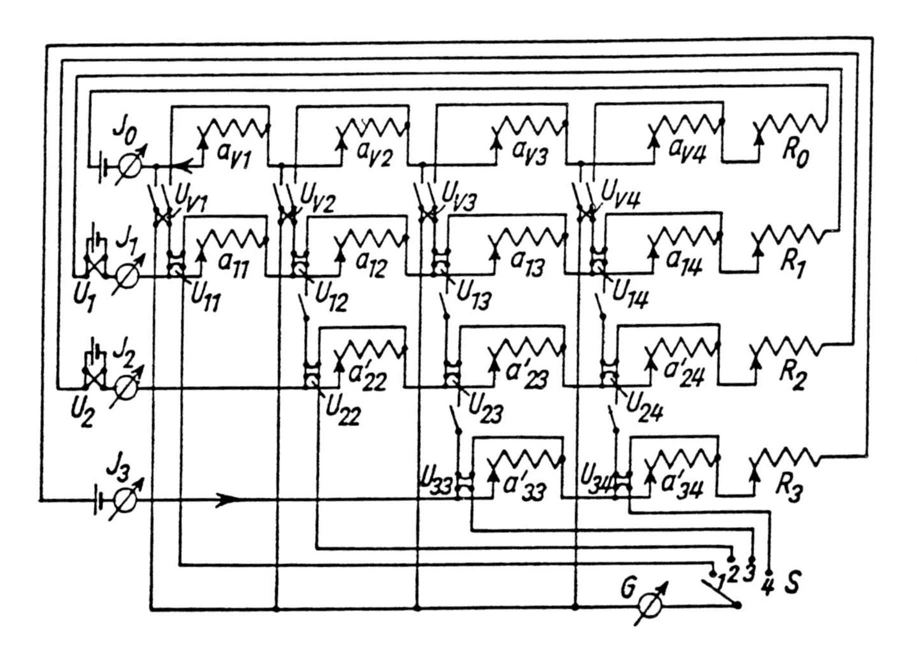

*Figure 1. Elementary diagram of the machine, as it was published by Cauer in 1935 (*Elektrische Maschinen *... p. 150). Notice the diagrammatic approach is highly standardized, but in a way that provides an abstract picture of the technical artifact which makes aspects visible one cannot see directly. However, engineering diagrams like this cannot elucidate the differences among the machines discussed here.*

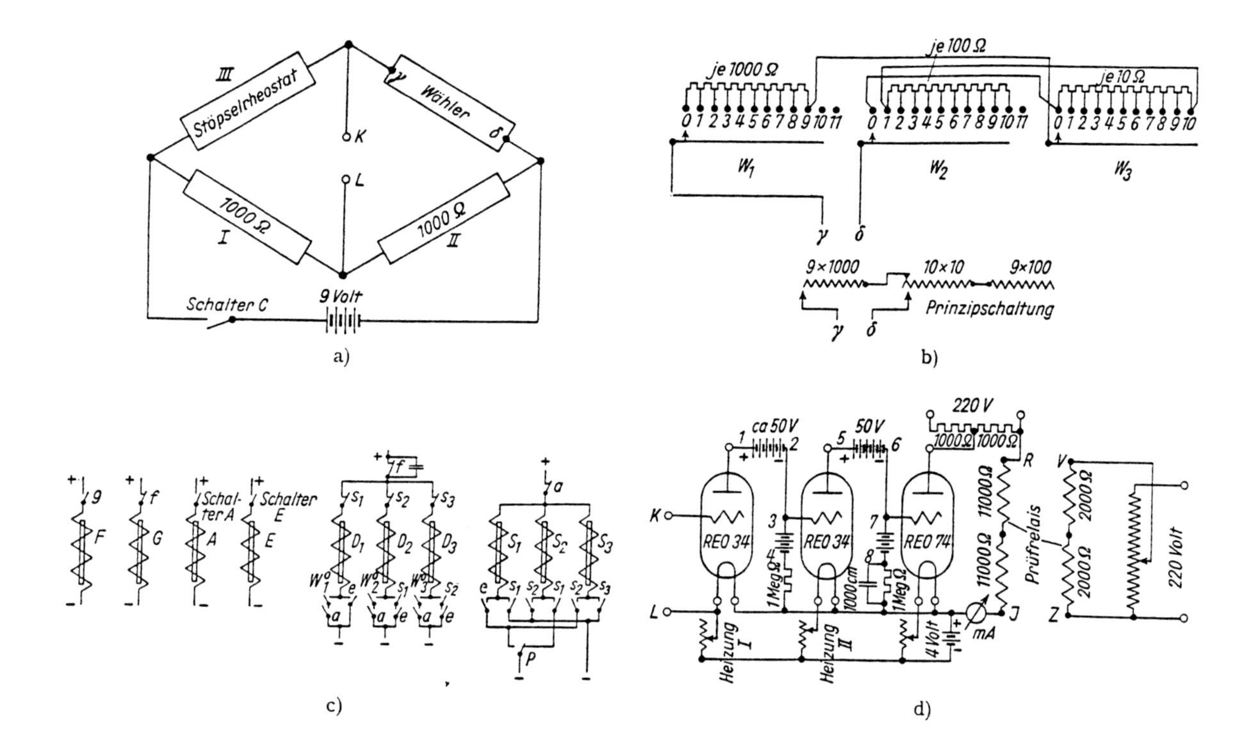

Figure 2. Diagram of the components of the automatic version: a) Wheatstone bridge circuit. b) four resistance arms of the bridge (three fixed and one variable), c) magnets and switches controlling the variable resistance), d) dc amplifier with polarized relay. (Cauer, Elektrische Maschinen…, *p. 150)*

from his publication: the coefficients were set by means of decadic resistors with 1000, 100, and 10 Ohms. Cauer wrote that a machine for ten equations would be feasible under the same principle and that it would be as expensive as Mallock's machine (below). To make it automatic, the galvanometer was replaced by a polarized relay connected with a d.c.-voltage tube amplifier. In the bridge diagram (a) resistors I and II are fixed. Resistor III is measured by reading the decade resistor gd, which is automatically set by three rotary switches W_1, W_2 W_3. The automatic setting of one coefficient took 3 seconds, Cauer reported.

Cauer emphasized that the elimination procedure could be arranged so that the capacity of the machine would be never exceeded, and also that from the calculated coefficients the value of the determinant and the errors could be evaluated. But this could not be done automatically, and the solution of a complete system of linear equations had to be complimented by separate manual exercises.

At that time Cauer was a Privatdozent for Angewandte Mathematik at the mathematical institute of the University of Göttingen. With his habilation thesis completed in 1928, he had had the opportunity to obtain an assistantship. This had been created under an agreement between the Rockefeller Foundation and the Prussian authorities, by which the famous mathematical institute of David Hilbert and Felix Klein got a new building and research positions. In the same year Richard Courant was named director of the institute. From Constance Reid's biography of

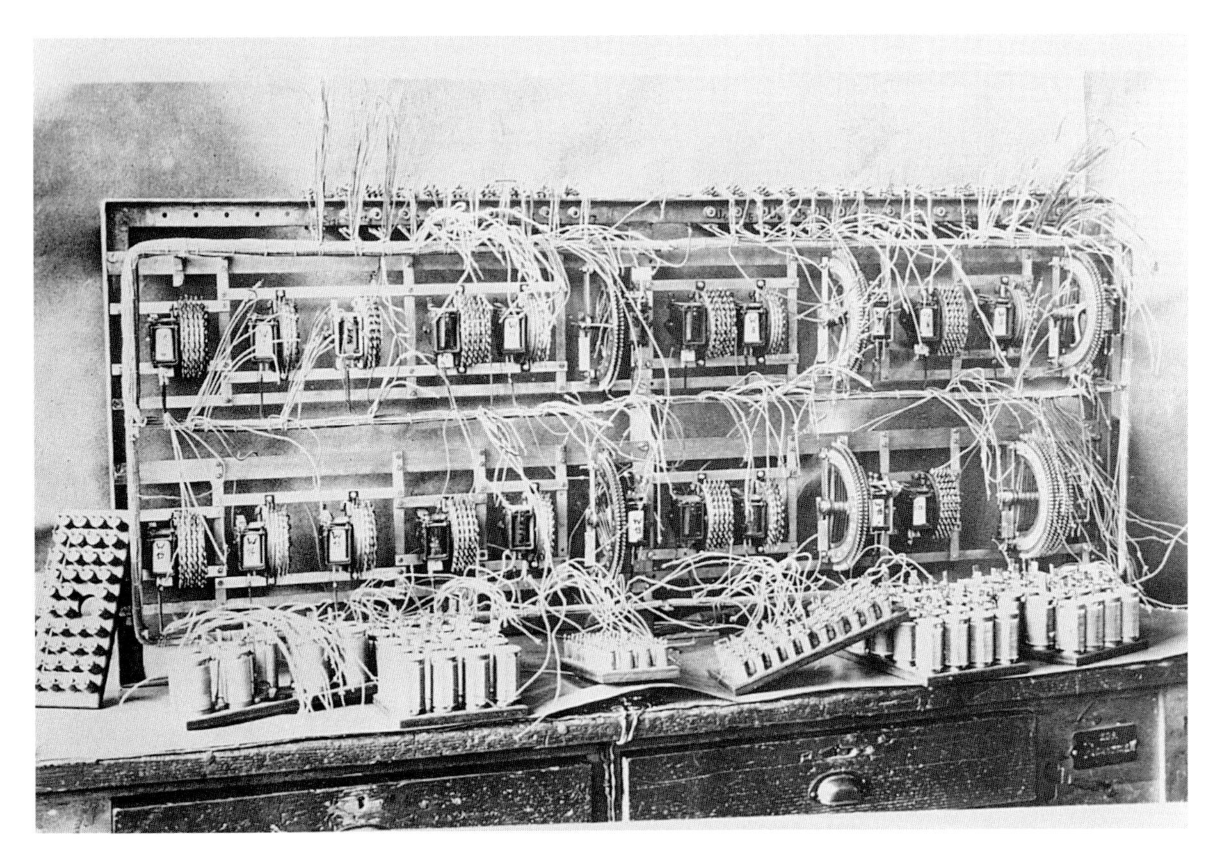

Figure 3. Photograph of the machine, probably in 1930. Allegedly the picture shows it in the entrance hall at the Mathematical Institute before Cauer left for the USA. He may have taken the picture to show to his American sponsors. (Cauer Papers)

Courant we get an impression of what a complicated man he was, and also how he managed the institute.[15] Nonetheless in 1928 Cauer had been able to push his Habilitation, for the subject Angewandte Mathematik.

Cauer used Courant's good relations with the Rockefeller Foundation to get a fellowship for one year, starting in September 1930, at Vannevar Bush's institute at the Massachusetts Institute of Technology (MIT). The correspondence with Bush just before Cauer's departure includes a discussion of the mathematical device. In one letter Bush mentions that he had discussed the subject with Norbert Wiener, but ultimately he was skeptical of Cauer's plans.[16] In Cauer's official report of the stay the calculating machine project is not mentioned. Nevertheless one can read that he had inspected several existing machines at MIT and at other institutions in the USA.[17]

How well Cauer and his wife mixed with others at MIT is unclear. Norbert Wiener, in his memoirs, mentions them only once.[18] But Cauer certainly learned that problems were handled differently in America. During the thirties Bush and his colleagues planned and built a whole series of important calculating machines. Bush, who was international recognized, broke through narrow academic rules in a way that was not possible in Germany, and he still had not reached the pinnacle of his career. In contrast, three years after his return to Germany, Cauer had to stop his project without any chance of resumption. This was due at least in part to the fact that the new telecommunications theories had not found an adequate place at the Technische Hochschule or at the universities. Academic institutions did not welcome a mathematical working engineer like Cauer. As a consequence, his academic career ran aground and he had to take up work as an engineer in an industrial laboratory and his influence on academical theories of engineering was modest.[19] When in 1941 the first volume of his major work 'Theorie der linearen Wechselstromschaltungen'[20] was published, he still held the position of head of Mix und Genest's research laboratory and gave lectures at the Technische Hochschule.

Cauer's difficulties in achieving an academic appointment at the university were also political. His best chance came in 1933. His fellow teacher at Göttingen, the statistics professor Felix Bernstein, was in the United States when Hitler came to power and he decided not to return. Cauer, who was married and had two children, applied for the vacant position. He was unsucessful, apparently because he was not a member of the Nazi party and he wasn't supported by the strong Nazi group at the university.[21]

Frustrated in his pursuit of an academic career, at the end of 1933 Cauer wrote letters to 18 scientific institutions and industrial companies, inquiring into their needs for solving linear equations. He posed seven questions, including the number of such problems which were dealt with in a year, the time needed for them, the number of unknowns and figures, and the type of instruments used. Most of the companies were concerned with static problems.[22] The solution of a system with 4 or 5 unknowns took anywhere from 45 minutes to 4 hours, 6 to 10 unknowns could take from 6 to 20 hours. An airship company (probably Zeppelin) had dealt with problems up to 24 unknowns and was interested in systems with 38 unknowns. There were photogrammetry problems with 68 unknowns, and an astronomical institute had no less than 360. The required precision 'in the engineering cases', as Cauer expressed it, would be no more than one per thousand. The demand would grow if one could calculate faster, thought Cauer.

In 1934–35 Cauer abandoned not only his plans for a calculating office and the completion of his mathematical machine. He left the university and joined Fieseler Aircraft at Kassel not far from Göttingen. In 1936 he was given a position as head of the research laboratory at Mix und Genest, and the family returned to Berlin.

Clark Leonard Hull's Machine

Cauer was particularly impressed by the design and the application possibilities of the calculating machine of the American psychologist Clark Leonard Hull (1884–1952). This machine was nearly unknown in

Germany and Europe. The fact that a psychologist built a self-designed semi-automatic calculating machine challenges the widespread impression that only people in commerce and engineering designed calculators. Some decades later experimental psychology would become an important part of the computer-using community. Nevertheless a career like Hull's, which illustrates a very pragmatic attitude, can hardly be imagined in Europe. It is also hard to imagine a greater contrast between the self-made scientist Hull and that of the academically-trained Cauer. These differences also found expression in their machines.

Born on a farm in New York state Hull attended a one-room school and later passed a teacher's examination. Because of illness he had to give up his planned career as a mining engineer and decided to become a psychologist. In 1914 he became a teaching assistant at the University of Wisconsin and wrote a doctoral dissertation on Chinese ideographs, which was soon accepted as a standard work on learning theory. In 1918 he was appointed instructor and was given responsibility for the course in experimental psychology.[23] Here he took over the work on tests and measurements in the Psychology Department and became interested in aptitude tests.[24] The interpretation of the tests depended on an extensive use of correlation coefficients, which the limited capacity of the then-available calculators made a tedious and error-prone process. So in February of 1921 he began to design a special machine to do the job automatically. He developed mechanical reproductions of the formulas of Thurstone for the Pearson coefficient of correlation and of Rumel for standard deviation.[25]

Hull planned all of the processing and mechanical details. In April 1923 the construction of the machine was commenced and it was presented, in a provisional state, to the Madison meeting of the American Psychological Association in December that same year. It was sufficiently perfected in the late summer 1924 to solve multiple regression equations automatically as well as to do practical correlation work on a large scale, with about 150 coefficients being computed. In December 1924 he demonstrated the machine before the Washington meeting of the American Psychological Association both as an automatic calculator of multiple regression equations and as an automatic correlation calculating machine. From April until December 1925 it computed over a thousand correlation coefficients and standard deviations.

In 1923 Hull had considered 'as a somewhat Utopian speculation' that some day a computing apparatus might be constructed which would solve multiple regression equations automatically, and thus yield mechanically, and cheaply, series of aptitude forecasts for vocational guidance.[26] But soon after, while working on the final details of the machine's design, which was planned only as an automatic correlation calculator, he hit upon the basic principles of such an aptitude-forcasting device. He found similarities of principles between the two machines such that the correlation machine as it

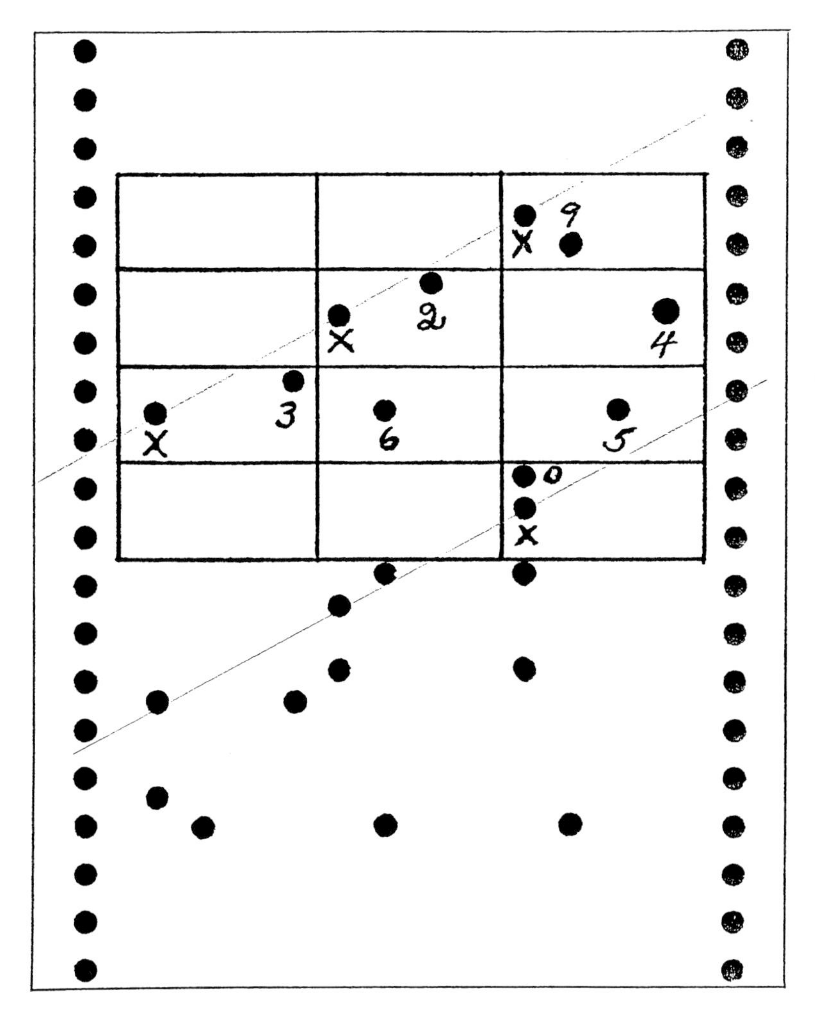

Figure 4. A sample strip of perforated tape (10 cm wide). Holes within the rectangles represent individual digits. Thus the first number is 9, the second 24, the third 365, the fourth zero. Holes at the left edge of the box for the first digit of each number control shifting from one number to the next. Successive columns of data to be correlated follow one after the other on the tape, a single blank space being interposed between each. (Hull, 'An Automatic Correlation Calculating Machine,' p. 526. The same picture was published by Hull in Aptitude Testing, *p. 488.)*

had been finally constructed served both purposes equally well. It could be used in two modes: one for the calculation of the arithmetic mean, the standard deviation and the coefficient of correlation; and one for the continuous calculation of the multiple regression equations, especially for making wholesale multiple aptitude predictions for purposes of vocational guidance and selection. No question that such a seemingly flexible machine was interesting also for Cauer's problems. Particularly promising, while the machine was doing its calculations, the attention of the operator apparently was not required at all. On several occasions, as Hull reported, he himself had started the machine on a long and difficult column of computations, then he had locked the laboratory and gone out to lunch.

The electrically-driven machine was essentially a products-sum calculator which rendered any special correlation tables or preliminiary hand computations superfluous. It calculated automatically the values required by the formulae, one after the other. The formulae were then ready for

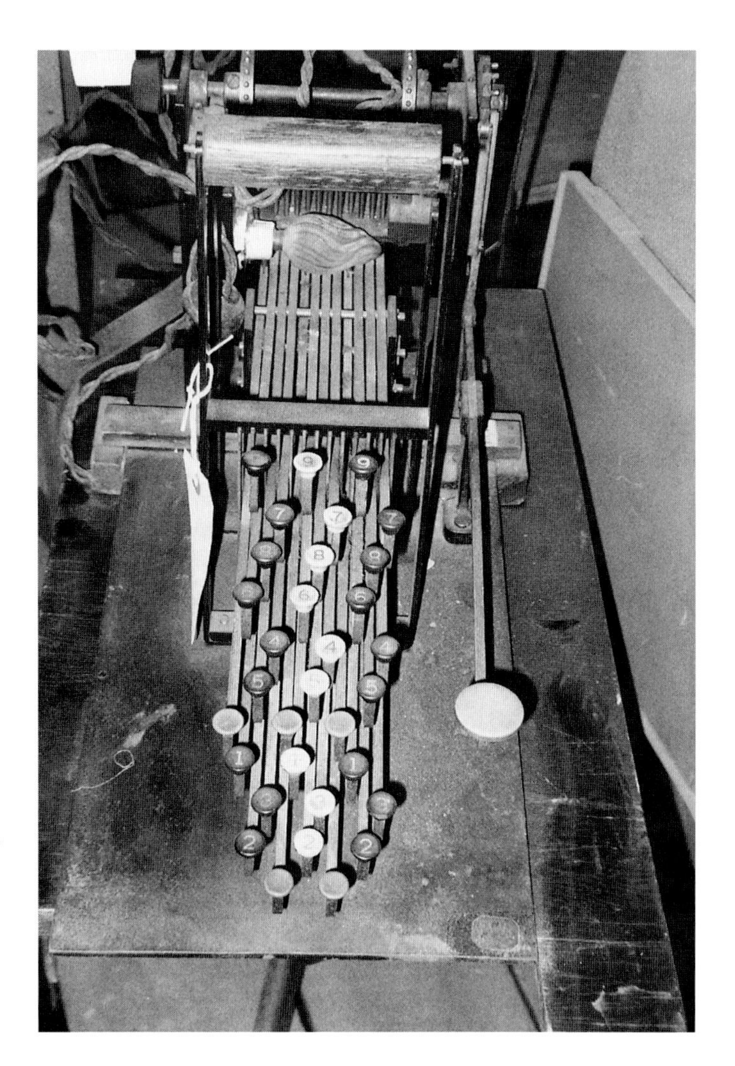

Figure 5. The special punch machine without tape. Notice the toothed wheels for moving the perforated tape, the plates to contol the punched numbers, and the special bulb which is able to provide a diffuse light for all positions of the plate. (Photo taken by H. Petzold in the NMAH, 1997.)

the final solution. The coded data then were punched into a strip of tough kraft paper, 0.07 inches thick an 4 inches wide, which was perforated on the edges somewhat like movie film. These edge perforations engaged the teeth of spockets on recorder and calculator in turn, insuring the precise movement of the paper through each. Numbers up to 999 could be recorded. A Veeder counter showed automatically the number of the item recorded at any given stage of the process. Cauer's equivalent solution with a rotary switch seems smarter. But Hull's punched tape also stored the data which was not possible with Cauer's device. Since the same regression equation was likely to be used over and over again the data for it were punched into a more permanent material such as a thin metal band, with 24 numbers on one foot of tape. The successive columns of data were correlated following each other on the tape, a single blank space being interposed between each.

Hull also considered how to correct errors. He placed a transparent number plate on the recorder over which the paper record was drawn as it feeded through the apparatus. The number corresponding to each perforation could be seen distinctly through the perforation itself. If an error was found, a small square of gummed parcel-wrapping tape could be pasted over the faulty perforation and the correct hole was repunched by the recorder.

The machine was built on a steel table, 26 inches by 32 inches. It was driven by a 1/4 HP electrical motor. Both at the multiplier and multiplicand there were steel fingers arranged in a pattern, one finger for each possible digit. These fingers periodically descended upon the perforated data strip. Those fingers which fell upon perforations passed through and thus directed the action of the machine very much as the hands of an operator might do by operating keys. When a multiplication had been completed the strips were moved to the next number. Squaring operations were performed simply by running duplicate data strips through the machine in parallel, one through the multiplier and the other one through the multiplicand. The action of the machine was purely mechanical, electricity being used only to drive the motor.

All that the operator was required to do was place the strip containing the numbers of the first column in one position, the strip containing the numbers of the second column in the other position and then press the starter. The machine then, automatically multiplied each pair of numbers one after the other continuously, adding up the products as it went along. When the machine reached the bottom of the columns, whatever their length, it stopped automatically and *(A × B) could be read from a dial.

Part of the funds for the machine came from the University of Wisconsin, a larger part from the Committee on Scientific Problems of Human Migration of the National Research Council. The machine was built by the mechanic Harold C. Kidder from the university in cooperation with the Chief Mechanician, O. E. Romare.

By December 1925 two machines had been built, the first one for the Wisconsin psychological laboratory, the second, an improved model, for the National Research Council. Hull then offered replicas of the calculator and the two auxiliary machines for about $ 1200.[27] As Hull reported, the establishment of a central correlation bureau had been suggested, since many institutions needing correlation work did not have enough work to make the purchase of a machine worthwhile. At the end of 1925 Hull wrote 'It seems likely that such a bureau provided with one or two of the machines … together with other modern statistical aids and a specially trained person, could do a good share of the correlation work of the United States and with a promptitude, economy and accuracy previously unknown.'[28] Today one machine is situated in the Smithsonian's National Museum of American History.[29]

Figure 6. The calculating machine in the depository of the NMAH. (Photo taken by H. Petzold, 1997.)

Hull actively promoted the machine.[30] In a textbook on aptitude testing, published in 1928, he described the rapid development of psychological testing during the previous years.[31] He was obviously proud that the former primitive methods had been replaced by scientific ones. Naive approaches to relationships between test and aptitude, gave way to better theories and clearer expectations. 'In a word, aptitude testing, like medicine and engineering, is ceasing to be a job for amateurs and is becoming the work of technically trained professionals.' (V). All of which was made possible by the existence of the machine. This is exactly the kind of argument that would have appealed to Cauer: his machine should in like fashion provide a new basis for the whole of mathematical engineering.

In 1925 Hull had argued in several papers as a practitioner: his machine would eliminate the drudgery and the persistent arithmetical errors from calculations of standard deviations and Pearson product-moment coefficients of correlation. In his 1928 book the arguments were more fundamental. He referred to Plato's argument in the *Republic* that everybody should be given his place in the state, and particularly in the army, in accordance with his abilities. To do so, it was necessary to detect those abilities. Plato's utopian dream, 'each man work at a single occupation in accordance with his natural gifts' was not possible without psychological tests.[32] Even if Cauer had comparable ideas he would not have dared publish them.

Hull saw the development of his calculator as 'another system of making aptitude predictions from forecasting formulae' and as an integral part of a comprehensive program of vocational education, which he

Figure 7. The detail shows arrangement of the reading fingers. One can see the massive design of the parts of the machine, which is stronger than any recent calculating machines. (Photo taken by H. Petzold, 1997.)

sketched for the first time in 1923. He called for the construction of one universal battery of tests which would sample all important aptitudes. The battery would cover some 30 or 40 different elements and its execution should take one day or more. Based on this battery separate formulae could be constructed for prediction. Forty or fifty different equations would have to be solved. All in all there would have to be about 1500 multiplications where the products had to be added. In cases like this only the machine offered freedom from arithmetical errors and other mistakes.[33]

Undoubtedly Cauer was impressed by Hull's sophisticated and obviously successful machine, especially since success in this case meant significant academic acknowledgment. He hoped to get a Hull machine to Courant's institute, and on 12 July 1929 he wrote to Hull in Wisconsin saying that he had read Hull's 1925 article in the Journal of the American Statistical Association 'mit großem Interesse', and that he would suppose that Hull's machine could be used 'auch zur automatischen Bewältigung mancher anderer zeitraubender Rechnungen, wie z.B. die Herstellung der Normalgleichungen bei Ausgleichsrechnungen ... Doch wäre es dafür erwünscht, eine Maschine mit größerer Stellenzahl zu besitzen.' He continued 'Wegen der außerordentlichen Wichtigkeit einer derartigen Maschine für numerische Rechnungen besteht der Plan, für das Göttinger Mathematische Institut eine solche Maschine anzuschaffen oder zu bauen' and 'falls sich der Preis tatsächlich in den von Ihnen angegebenen mäßigen Grenzen bewegt'. He apparently got no answer, but he clearly continued to be interested. He was responsible for a footnote mentioning Hull's machine which appeared in a publication by the mathematician Theodor Zech in March 1929.[34] And in a letter dated 3 January 1930 Cauer asked Bush if he could say anything about the usefulness of Hull's machine.[35] During his stay in the United States Cauer met Hull and saw the machine at the Institute of Human Relations at the Yale University. Unfortunately there is neither a record of his impressions at the time nor an indication in the Cauer papers if he made any further effort to get a Hull machine to Göttingen.

The 'Columbia Machine' of IBM

Cauer was familiar with the punched card system. He inspired Theodor Zech to consider the use of this system for harmonic analysis. Undoubtedly it was also Cauer who inspired Zech to collect data associated with renting and operating a machine. The data came from the Vulkan dockyard in Hamburg and from the Dehomag company.

During his stay in the USA, Cauer saw the new tabulating machine which had recently been developed at IBM. At this time IBM had been

Figure 8. The Columbia Machine—with Benjamin Wood at his calculation center in 1930 (Archives of the NMAH)

Figure 9 The same machine in 1997 withPeggy Kidwell at the Smithsonian. (Photo taken by H. Petzold, 1977)

building and marketing punched-card machines for three decades. In retrospect there is no question that the development and the success of the punched card system by IBM and some other smaller companies paved one of the different ways that would lead to the computer. The first president of the company and creator of the name IBM, Thomas J. Watson, executed the strategical switches. Some of them can be designated as historic not only for the company but generally for the development of information technology. One turning point in the history of the company occurred at the same time that Cauer was planning his equation solver in Germany.

In 1928 army psychologist Benjamin Dekalbe Wood became a professor of Collegiate Educational Research at Columbia University. Son of an itinerant cattle rancher in Brownsville, Texas, Wood, like Hull but unlike Cauer, did not grow up in an academic environment. His school education had been short and he was in most fields self-taught. As a student at the University of Texas with no high school credits, he had studied Plato and Aristotle with an obviously naive and irreverent spontaneity, and was impressed by Malthus.[36]

The chronicler of IBM and biographer of Thomas J. Watson, William Rodgers, tells us that Wood left university with a Batchelor's degree convinced that, in the words of H. G. Wells, 'civilization is a race between education and catastrophe' and was determined to intervene. After his military service, he attended Columbia University and for his doctoral dissertation applied E. L. Thorndyke's thesis that 'whatever exists at all, exists in some amount' to the measure of human intelligence. As an assistant to the director of the Columbia College, Herbert Edwin Hawkes, he advised the students and found himself confronted with the question of how their future development could be assessed during the term of their vocational education. His methods of tests and valuations attracted some attention. Wood was also in contact with John Dewey and his Progressive Education Association.

With Hawkes' support Wood got substantial funds from the Carnegie Foundation,[37] the Commonwealth Fund, and the Educational Board. His investigations also made the authorities of the education bureaucracy listen attentively. When the interpretation of test results exceeded the capacity of current processing, Wood wrote to the chief executives of ten corporations in the equipment business and asked for support. It is part of the IBM legend that its autocratic president was the only one who answered, and that Wood convinced him in a long dialogue, after which Watson supported Wood with a calculation center based on the punched card system.

Alledgedly Wood convinced Watson that the limits were defined by the productivity of the machines. The goal was to work with the speed of light. As Rodgers reports, Watson put pressure to his technical staff to achieve this goal. He rejected their arguments that the possibilities of technology were limited.[38] Even if the whole story, written forty years later, was fabricated as propaganda, there is no doubt that during the following decades Watson was determined to open up new markets in the sciences through new modifications to the standard machines.[39] Watson knew that punched card technology would go far beyond the hitherto-existing markets of IBM, indeed it had no limits at all. This attitude is in stark contrast to that which surrounded Cauer, or even Hull and Wood.

Figure 10. The front of the Columbia Machine is unique not only in the way the input-output technology is treated, but also in the typical prewar IBM design: completely black with golden ornamental lines. (Photo taken by H. Petzold, 1997.)

In 1929 the IBM president installed a calculation center with punched card machines for Wood at Columbia University. From Watson's point of view it would play the role of a model and a playground, where protagonists' activities and the results they achieved could be observed. The first important modification of a standard machine, incorporating the suggestions of the scientists around Wood, was the so called Columbia tabulating machine. It was a modification and enlargement of then new type IV tabulating machine. Engineers James W. Bryce, George F. Daly and Gunne Lowkrantz in the IBM plant at Endicott, New York, built a machine with two counters for transfer cycles and ten counters with ten figures each. The machine had the ability to transfer numbers between the different counters. However this still was far from being a computer; the machine could only add. IBM marked the new claim with a patent where the text particularly mentioned scientific applications.

This machine arrived at Wood's calculating office in December 1929 and is known as the 'Columbia Machine' or 'Ben Wood's Machine.'[40] It remained unique and has also found its place in the National Museum of American History.[41] The 'Columbia Machine' was IBM's first step in a line of development which lead in 1935 to its commercial test scoring machine, the IBM 805. In Germany the Dehomag developed at the same time the tabulator machine D11 without any contacts to scientific institutions or individuals.[42]

I am convinced that this was the 'new machine' which Cauer saw during his visit and is mentioned in his report. But we do not have any more direct information about his thoughts. He must have been impressed by IBM's interest in applications of the punched card system for scientific calculations, particularly correlation calculations, and also by IBM's readiness to modify the conventional machine for a new scientific market. But he must also have noticed the quantities of data which were needed for these sociological and statistical evaluations, which exceeded the quantity of the coefficients of Cauer's equations by many times. He must also have recognized the particular limitations of the punched card system. Most impressive to Cauer would have been the high standard of reliability which was set by the IBM devices, and which was impossible to reach with his own machine in the contemporary Göttingen situation.

It is interesting to speculate what had been happened if Cauer had been the one to approach Watson with his method of solving linear equations. Chances are that he would not have been as successful. Partly it would have been a personality problem: Cauer did not have the self assurance possessed by Wood. But Cauer also would not have had the vision of a market with hundreds and thousands of equations where millions of punched cards would be needed—and where the small errors allowed by his system would have become intolerable.

J. B. Wilbur's Simultaneous Calculator

In the spring of 1935 Cauer became aware that at MIT J. B. Wilbur was building a machine that could solve a system of nine equations with nine unknowns, working on a 'purely mechanical-kinematical principle'. It functioned on what Bush had described as a more precise version of the well-known principle developed fifty years before by Lord Kelvin.[43,44] Wilbur did not claim to be the inventor of the machine; he described his role in the project as a 'clearing house for the ideas of those who work with him.' Obviously there had been many such ideas. He thanked Bush who was then the Vice President and Dean of Engineering at MIT. The technical realization of the machine was made possible by the Singer Sewing Machine Company and its President Sir Douglas Alexander. The head of the project was Professor Charles B. Breed, Head of the Civil Engineering Department at MIT.

Certainly Cauer had not forgotten his exchange of ideas with Bush from 1930/31 when he revealed the details of his own concept. Now he had to consider the Wilbur equation-solver, based on another principle, which was not new and which Cauer had rejected six years before.

Wilbur's published description of the machine was written in December 1936, too soon to include a critical report on its practical use. It leaves open the question of why they had decided on Kelvin's principle (an approach that Cauer had explicitly rejected), which had some problems and inevitably made the machine unwieldy in size. In 1934 a rough prototype had been sufficiently successful in solving systems of linear equations with real coefficients so that a decision was made to build a larger machine. This was essentially complete by 1935, and in 1936 it was used for solving systems of nine equations. Wilbur emphasized that it could be also used for calculations of the unknowns of more equations if some restrictions on the form could be made. He reported that 'theoretically' this could be done without restrictions with modified versions of the machine.

The optimistic generality of Wilbur's statements can be contrasted to the skepticism expressed by Bush in his letter to Cauer in 1930. In 1936 Wilbur wrote confidently that the machines would be helpful for the progress of technology and research.[45] But apparently it was characteristic of the times that any real market analyses would be left to the financiers and was not a concern of the scientist-inventors.

Wilbur's machine consisted of a heavy steel case with ten steel plates arranged to swivel inside. Nine plates corresponded to the nine unknowns, the tenth to the constants. The solution of nine equations with nine unknowns, to a precision of three digits, took between one and three hours. Although this was better than working with a desk calculator (where calculation of eight equations with eight unknowns took on the order of eight hours) Wilbur hoped that with practice the time could be shortened. For precision problems Wilbur mentioned only that the time

Figure 11. Wilbur sitting before the Simultaneous Equation Solver, setting a coefficient. (Photo MIT Museum)

taken depended on the type of the equations. He believed that for ordinary systems of equations the error would not exceed 1% of the largest unknown. In most cases precision would be greater. The machine was suited for stepwise approximation, which made possible precision to any degree desired. Under favorable conditions, a system with 18 equations had been solved in 7 or 8 hours with a precision of four or five digits. The same calculation done with a conventional desk calculator needed 32 hours. A greater number of unknowns would increase the time saved significantly.

Wilbur wrote that research and development with this type of mechanism should be continued, stating confidently that a machine for the direct solution of a greater number of equations was possible. However, even on the first machine, small modifications were necessary. One technologically weak point, which Wilbur already had noticed, was the type of steel tape used which was expected to lose its flexibility and tension. The technical skill required to handle the machine needed to be improved. And he expected reading accuracy to be increased. In an improved bigger machine time could be saved by development of an automatic frame for setting coefficients and constants (another indication of the move towards automation).

The only practical use of Wilbur's machine was by the prominent Harvard economist Wassily W. Leontief from Harvard University.[46] Leontief's original calculation of a national economic 42-sector input-output model required about 30 million multiplications. He simplified the data into a 10-sector grid, but even that would have required 450,000 multiplications or, as he reckoned, two years at 120 multiplications per hour. Leontief recalled in an interview in 1969, 'You could really change the coefficients slightly by simply sitting on the frames, and if they did not give too much this meant that the solution was relatively stable.'[47]

Wilbur would never build an improved machine. He decided to make his career as a professor of civil engineering at MIT and not as a designer of mathematical machines.[48]

Mallock's Machine

Our last machine is that of R. R. M. Mallock in England. Both in form and function it resembled Cauer's more than any other. Its story suggests a similarity of approaches between Germany and England (more so than with the United States) and also how Cauer's machine might have evolved. The emergence of Mallock's machine, and a new publication by Bush,[49] gave Cauer reason to believe that his ideas had been confirmed, and he wrote a paper claiming priority.

Rawlyn Richard Maconchy Mallock was born in 1885 in South England where his father represented the community of Torquay in Parliament. Mallock took an engineering apprenticeship in Manchester

Figure 12. The detail shows one of the plates with some of its nearly 1000 pulleys leading the steel tape. Above and below some of them can be shifted horizontally. Notice the controlling knobs at the plate. (Photo MIT Museum)

in 1906, and 1908, after he had completed part one of the mathematical and mechanical sciences tripos at Trinity College, Cambridge. He went to Canada for some time and satisfied his war service with the armaments company Armstrong Whithworth. After the First World War he was an electrical expert at HMS Vernon, a research institution of the British Admiralty. At the end of the 1920s he returned to Cambridge and worked there until his retirement in 1937. He died in 1959.[50]

Mallock gave the same reasons for his project as Cauer and the others: 'In connection with many problems of engineering and physics it is necessary to solve sets of linear algebraic simultaneous equations involving a large number of unknowns; for instance in the determination of secondary stresses in bridges and other structures sets of equations involving from ten to twenty, or even more, unknowns may occur and the labour involved in the solution when the number of unknowns is more than about six is very great.'[51]

In 1931, almost at the same time as Cauer, Mallock constructed his prototype, which solved systems with 6 linear equations with a mean error of 0.4%. The mean error had been reduced to 0.1% in the following machine, built in 1933.[52] In Mallock's eyes this was enough for the solution of engineering problems. Like Cauer, Mallock designed two additional frames which could set the rotary switches automatically by means of several relay circuits.

Mallock made a contract with the Cambridge Instrument Company, to which he granted his patent rights for the construction of future machines. Charles Darwin, the chairman of the Company and later head of the National Physical Laboratory in Teddington, mentioned the contract in his report for the Royal Society, 'in an endeavour to create a real market for this machine'[53] At the beginning Mallock contributed £200, with the remaining cost of construction to be payed by the company. The expected profit was to be divided according to this ratio until Mallock had received 400 Pounds.[54]

In 1933 Mallock directed construction of a machine at the Cambridge Instrument Company. After its completion it was taken to the Cambridge University Engineering Laboratories, where Mallock was employed. Later, in 1937, it was bought by the university's Mathematical Laboratory (for £1,750; Croarken is convinced that the company lost on the deal, all things considered) where one of its users was the computer pioneer Maurice Wilkes.[55] In 1933, Darwin presented the machine to the Royal Society, and published a comprehensive report in its *Proceedings*.[56] A further report was published in the widely-read technical journal *Engineering*.[57] In addition, professional circles were informed of the

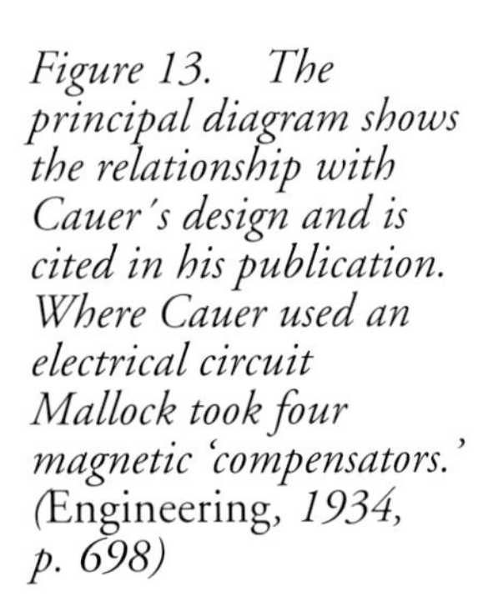

Figure 13. The principal diagram shows the relationship with Cauer's design and is cited in his publication. Where Cauer used an electrical circuit Mallock took four magnetic 'compensators.' (Engineering, *1934, p. 698)*

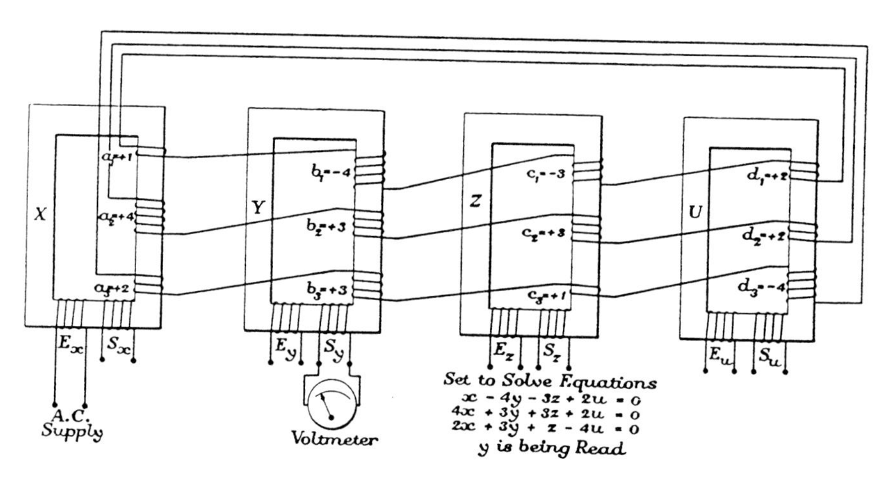

machine at the Royal Society Conversazione in May 1933. All of which illustrates that Mallock had the kind of support that Cauer lacked.

As a consequence, the Mallock machine commanded a certain degree of interest for about a decade. But eclipsing both Mallock and Cauer was Bush, who described (and promoted) his Differential Analyzer in 1934 at the International Congress of Applied Mechanics in Cambridge, England.[58] Quite in contrast to Cauer's stay in United States, which was noticed only by ten or twenty fellows, Bush's appearance was a high-profile affair which can be seen as part of a new scientific-political initiative on the part of the United States, supported (as Cauer's trip had been) by the Rockefeller Foundation. Over the succeeding years differential analyzers—often called 'Bush machines'—were constructed in several European countries, this putting the stamp of mechanical analogue technology on mathematical machines until the end of the 1940s.

There were reasons for the different reactions of the scientific and mathematical communities to the machines of Mallock and Bush. For one thing, the Differential Analyzer had had several forerunners both in terms of machines and practice ranging back to the beginning of the 1920s. Moreover, it could solve differential equations, giving it a clear advantage over a machine that could only solve less spectacular linear equations.

The different ways of working may also have played a part: The Differential Analyzer, as an analog machine, was fed with curves, was operated with curves and produced curves. And most scientists and all engineers were taught to visualize their problems in terms of curves. Mallock's machine (like Cauer's) had a digital link to the omnipresent classical, industry-made desk calculators. But to a certain extent it lost this advantage because of its hybrid nature; one could say that its results were measured rather than calculated or reckoned.

Mary Croarken, who has done extensive research on Mallock and his machine, states that the prominent figure of Comrie was never particularly interested in Mallock's machine and that he never studied it intensively. The reports in *Nature*[59] and in other journals confined themselves to descriptions of the technical function and failed to evaluate its reliability or its relevance to the scientific and engineering community. Furthermore, Mallock, like Cauer, does not seem to have had a prophet's temperament.

Mallock's machine is the only one of this group which was used—not intensively, but several times—and where the experiences of that use are known. As the mathematician A. C. Aitken reported in *Nature*,[60] the setting of the switches and the plugging of the wires for the calculation of a system with 6 equations, took half an hour. When a solution had been determined it was possible to proceed to more approximations to get more precision without it being necessary to change the adjustments. But in most cases the machine was used only for the calculation of

Figure 14. Mallock's calculating machine (Photo C.S.I. Co.)

solutions with large numbers, since the smaller the numbers the less their precision.[61]

An engineer at the Royal Aircraft Establishment in Farnborough tried unsuccessfully to use the machine in 1936–37. During the war, the External Ballistics Department of the Ministry of Supply employed the Mallock machine occasionally when they were using the Mathematical Laboratory, as did people from de Havilland aircraft company, the Royal Establishment, and the National Physical Laboratory.

Croarken concludes that the machine was never really successful even though it was seen as 'a useful and reasonably accurate device.' Aside from some technological deficiencies, it apparently was unable to calculate 'ill-conditioned' equations, which are unavoidable in engineering applications. But there were also some uncontrollable feedback effects which made the machine hawl and pipe. The mathematicians Wilkes and Aitken did not have to deal with any 'ill-conditioned systems' and thus had good results. Presumably all of this would have been true of the Cauer machine, if the prototype had been really completed and used.

There was only a modest interest shown in acquiring a Mallock machine. In 1933 MIT asked, but they were unable to accept the tender of Cambridge Instrument. At the beginning of the war there was an inquiry from General Electric Corporation at Schenectady, and in 1944 another from the Royal Air Force. But because of pressure of competing work the company was not able to respond. In 1944, when Charles Darwin, in his new position as the director of the NPL, suggested the installation of a central calculation office for all governmental institutions, his plan included a Mallock machine. But in 1945 when this proposal became the starting point for the installation of extensive equipment at the NPL Mathematics Division, the Mallock machine was no longer included. After the war, in 1947, the Oscillations Sub-Committee of the Aeronautical Research Council tried to order a machine, but the company was still unable to comply and so gave the patent rights back to Mallock.[62]

Unlike Cauer, Mallock was able to dedicate the greater part of his life to the development of his machine. But because he did so, as Wilkes reports, its lack of commercial success left him a disappointed man. Only the installation of the machine in the Mathematical Laboratory gave him some pleasure. Wilkes noticed that the automatic setting frame was not completely developed in 1937 and concluded that the machine had therefore not been used at the institute at all. Nevertheless the successful pioneer who one decade later created one of the first modern electronic digital computers retrospectively summarized: 'However, it gave me my first introduction to the use of telephone relays in computing, or rather control, circuits and to some of the tricks that one can play with them.'[63]

Some Conclusions

The equation solvers described above are representative of spectrum of machines that were devised during the pre-war years. Although they were commercially unsuccessful, they can be seen as characterizing the historical-technological situation during a critical period, and we can understand the people in the focus of this paper as a community with parallel interests. Each member of this community, responding to common technical problems, was conditioned in his response by a variety of non-technical circumstances. Historians must consider both aspects. It seems that in Cauer's eyes the importance of these machines grew as his chances of becoming a professor declined. As we have seen, his prospects were influenced initially by an academic environment that was inimicable to his mathematical-engineering approach. And increasingly, they were influenced by the political situation in Germany. Mentioned above were Cauer's difficulties with the Nazi party. It is significant that this kind of problem extended even to Courant, who in spite of having been a particularly effective academic representative of the first German republic was despised by the Nazi academics at Göttingen. He had to leave in 1934 despite a new law that said that front-line soldiers from the First World War should be spared.

It is notable that around 1930 this new type of calculating machine for solving linear formulae appeared in two completely different fields of science: engineering, and psychology and social sciences. The former were given special prominence in Europe, the latter in the United States. There was some cross-over, of course. Punched card machines were used in the USA and also in Europe for census work, which can be interpreted as social-scientific problem solving. In 1933 IBM's German daughter Dehomag (Deutsche Hollerith Maschinen AG) offered this machine to the new Nationalsocialist system in Germany to help in effective implementation of its racial politics. For a short time the new situation at Germany seemed to offer a real possibility for empirical social politics on the base of the new technology. But there was no interest in an academic calculating center for social sciences. Instead, Dehomag found its market in the administration and military and industrial bureaucracies.

After 1934 the influence of Vannevar Bush, supported by the large foundations, was overwhelming. Bush not only popularized the technical possibilities of his big mathematical analog machine, he also demonstrated to the world that it could be financed, both in the United States and in other countries with help from the United States. It is difficult to decide if in Göttingen or elsewhere in Europe there was anyone with abilities comparable to those of Bush, but it is certain that nowhere else were there conditions comparable to those that he enjoyed.

In each of the cases presented, at least three concerns had to be addressed. First was to define a problem significantly different from any that which could be accomplished by contemporary desk calculators. Second was to select from numerous possibilities the technical method to be employed. Third was to gain financial support. This last was especially difficult in Europe, while in the United States at least two approaches (Bush and IBM) proved feasible.

There is one more aspect to these pre-computer years: the number of unknowns which the different scientists and engineers felt necessary to be determined was invariably low. One reason might be that they were accustomed to using calculators professionally and did not have experience with as many equations and unknowns as they would need today. Which means that these earlier artefacts might be considered, like telescopes, as capable of revealing a new land (of computing) but at the same time limiting its extent.

Notes

1. W. Aspray, *Computing Before Computers*, Iowa State University Press, Ames, Iowa 1990, 255.
2. Aspray, p. 179.
3. The most important textbook in German is F. A. Willers, *Mathematische Maschinen und Instrumente* (Berlin, 1951). Here one finds 871 cited papers on mathematical instruments and machines. It includes only some of the most important papers in non-German languages. Another textbook is W. Meyer zur Capellen, *Mathematische Instrumente* (Leipzig, 1944), where 310 papers are cited.
4. H. Petzold, 'Maschinen zur Lösung verwickelter mathematischer Probleme. Versuch einer historischen Ortsbestimmung der elektrischen Rechenmaschine Wilhelm Emil Cauers,' in W. Mathis and P. Noll (ed.), *Second ITG-Diskussionssitzung* 21–22, April 1995 (Berlin, 1995), pp. 267–282. This is a preliminary report. I express my thanks to the families of Emil Cauer and Wolfgang Mathis who entrusted Cauer's letters and papers to me, hereafter cited as Cauer-Papers.
5. E. Cauer, 'Familie, Bildung, Wissenschaft. Wilhelm Cauer 1900–1945,' in W. Mathis and P. Noll (ed.), *Second ITG* .... pp. 247–259.
6. As for Hamel, cf. O. Haupt, 'Nachruf auf Georg Hamel,' in *Jahrbuch der Akademie der Wissenschaft*

und Literatur (Wiesbaden, 1954), pp. 148–154, which includes a list of Hamel's publications; L. Prandtl, 'Georg Hamel siebzig Jahre,' in *Zeitschrift für Angewandte Mathematik und Mechanik*, 28 (1948), pp. 129–132; W. Kucharski, *Über Hamels Bedeutung für die Mechanik, in: Zeitschrift für Angewandte Mathematik und Mechanik*, 32 (1952), pp. 293–297. For more on Hamel's mathematical and ideological attitude see G. Hamel, 'Die Bedeutung der Mathematik in der heutigen Zeit,' in *Forschungen und Fortschritte*, 9 (1933), pp. 487–489.

7. The most important biograhical articles in Wilhelm Cauer are: E. Cauer and W. Mathis, 'Wilhelm Cauer (1900–1945),' *Archiv für Elektronik und Übertragungstechnik* 49 (1995), pp. 243–251; and E. Cauer, 'Familie, Bildung, Wissenschaft ....'
8. Hamel's most important books are *Elementare Mechanik* (Leipzig, 1912), and *Theoretische Mechanik* (Berlin, 1949).
9. Only occasionally can historians learn which precise instruments were used to calculate the unknowns of systems of linear equations. In his letter to Cauer from 22.1.1930 Vannevar Bush mentions that his 'ordinary method' of solving systems of linear equations was 'by means of a slide rule of the cylindrical type used together with an adding machine.' Cauer Papers.
10. Twenty-five years later Cauer's doctoral dissertation was accepted within the new engineering discipline of network synthesis. W. Mathis, 'Die Rezeption von Cauers Arbeiten. Aus dem Nachlaß von Wilhelm Cauer,' in *Second ITG ...*, pp. 289–294.
11. The earliest reference I have found is a patent office receipt dated 7 September 1928, Reichspatentamt Berlin. Cauer-Papers.
12. List of Cauer's patents in W. Cauer, *Theorie der linearen Wechselstromschaltungen*, Vol.1, 2nd Edition (Berlin, 1954), p. xvii.
13. W. Cauer, 'Elektrische Maschinen zur Auflösung von Systemen linearer Gleichungen,' *Elektrische Nachrichtentechnik* 12 (1935), 147–157. On page 150 Cauer mentions 'a suggestion' he had made 'already in 1928.'
14. The only description of the device is in W. Cauer, 'Elektrische Maschinen....'
15. C. Reid, *Richard Courant 1888–1972. Der Mathematiker als Zeitgenosse* (Berlin, 1979).
16. Correspondence with Bush between August 1929 and July 1930. Cauer-Papers.
17. W. Cauer, 'Report of my Activities as Fellow of the Rockefeller Foundation,' manuscript without date, obviously end of 1931. Cauer-Papers.
18. N. Wiener, *I am a Mathematician* (London, 1956), p. 142. Wiener wrote Cauer's prename incorrectly as 'Richard' instead of 'Wilhelm'—quite likely a confusion with. As far as I know, in the numerous different editions and translations of Wiener's memoirs this mistake was never corrected. Wiener continued, 'However, the scientist with whom I had the most interesting and profitable contacts was Eberhard Hopf.'
19. Cf. W. Mathis' analysis of the reception of Cauer's work. W. Mathis, 'Die Rezeption von Cauers Arbeiten. Aus dem Nachlaß von Wilhelm Cauer,' *Second ITG...*, 289–294.
20. W. Cauer, *Theorie der linearen Wechselstromschaltungen*, Vol. 1 (Berlin, 1941). Sec. ed. (Berlin, 1954).
21. A critical and conscientious report on the situation at Courant's institute has been written by N. Schappacher, 'Das Mathematische Insititut der Universität Göttingen, 1929–1950,' in H. Becker, et. al., *Die Universität Göttingen unter dem Nationalsozialismus. Das verdrängte Kapitel ihrer 250-jährigen Geschichte* (München, 1987), pp. 345–373.
22. W. Cauer, W., 'Betrifft Auflösung linearer Gleichungen.' Copies of this paper, which is dated from 1 October 1933, were obviously inclosed to Cauer's letters. Cauer Papers. The results of these project were published in: W. Cauer, 'Elektrische Maschinen ....'
23. R. R. Sears, Clark Leonard Hull, in J. A. Garraty (ed.), *Dictionary of American Biography*, suppl. 5 (New York, 1977), pp. 328–331.
24. C. L. Hull, *Aptitude Testing* (Yonkers-on-Hudson, Chicago, 1928). Cf. generally S. J. Gould, *The Mismeasure of Man* (New York, 1996).
25. C. L. Hull, 'An Automatic Correlation Calculating Machine,' *Journal of the American Statistical Association* 20 (1925), 522–531; C. L. Hull, 'An Automatic Machine for Making Multiple Aptitude Forecasts,' *Journal of Educational Psychology* 16 (1925), 593–598; L. L. Thurstone, *Psychological Bulletin* 14 (1917), 28; B. Ruml, *Psychological Bulletin* 13 (1916), 444. An arrangement on Hull's machine gives C. and R. Eames, C. and R., *A Computer Perspective. Background to the Computer Age*, New edition (Cambridge, London, 1990. (first ed. 1973), pp. 70, 71 and 89.
26. C. L. Hull, 'The Joint Yield of Teams of Tests,' *Journal Educational Psychology* 14 (1923), 405.
27. Obviously, this is the price which Cauer referred to.

28. Hull, 'An Automatic Correlation Calculating Machine...,' p. 531.
29. The machine had been received in 1955 with Accession Number 205,424 and Catalog Number 314,605 as 'Calculator, Dr. Clark Hull's coordination (sic!, must be 'correlation') machine.' In the archives of NMAH one can find along with correspondence related to acceptance of the machine a microfilm of several copy-books with Hull's daily notes and sketches of his stepwise design. Clark Hull Papers 1902–51, Reel 3. The original is at the archive of the Yale University. I express my great thanks to Peggy Kidwell for preparing and supporting my inspection of the machine at the depository outside of Washington D.C. and also for her reference to the microfilm.
30. I could not find any critical reports looking back upon the work with the machine. Unfortunately I could not investigate exhaustively and generally the history and the importance of Hull's psychological work. So I cannot refer to any critical comments of Hull's psychological fellows.
31. Hull, C. L., *Aptitude Testing*, p. v.
32. Hull, *Aptitude Testing*, pp. 5, 6.
33. Hull, *Aptitude Testing*, pp. 487ff.
34. T. Zech, 'Harmonische Analyse mit Hilfe des Lochkartenverfahrens,' *Zeitschrift für Angewandte Mathematik und Mechanik (ZAMM)* 9 (1929), 425–427.
35. Cauer Papers.
36. W. Rodgers, *Think. A Biography of the Watsons and IBM* (London, 1970), p. 134.
37. Some years later V. Bush became president of the Carnegie Foundation.
38. Rodgers, *Think...* p. 138.
39. E. W. Pugh, *Building IBM. Shaping an Industry and its Technology* (Cambridge, MA, London, 1995), p. 70.
40. Cf. F. W. Kistermann, 'The Way to the First Automatic Sequence-Controlled Caculator: The 1935 DEHOMAG D 11 Tabulator,' *IEEE Annals of the History of Computing* 17 (1995), 33–49, particularly 41.
41. I thank once more Peggy Kidwell for her assistance in my inspection of the machine.
42. Kistermann, 'The Way ....'
43. V. Bush, 'Recent Progress in Analysing Machines,' *Proceedings of the 4th International Congress for Applied Mechanics* (Cambridge, Eng., 1934), 3–23.
44. The Kelvin principle was used practically in numerous tide predicting machines. Cauer doubted still in 1935 that it was possible to control the propagation of the error in a purely digital automatical machine. The only possibility he saw was an analog tunable device. Cauer, 'Elektrische Maschinen..,' p. 154.
45. J. B. Wilbur, 'The Mechanical Solution of Simultaneous Equations,' *Journal of the Franklin Institute* 219 (1936), 715–724.
46. W. W. Leontief, 'Interrelation of Prices, Output, Savings, and Investment,' *The Review of Economic Statistics* 19 (1937), 109–132.
47. Cited in Eames, *A Computer Perespective ...*, p. 113.
48. I thank I. B. Cohen for this information. He had saved Wilbur's machine.
49. V. Bush, 'Structural Analysis by Electric Circuit Analogies,' *Journal of the Franklin Insitute* 217 (1934), 289–329.
50. I am greatly obliged to Mary G. Croarken who intrusted her unpublished manuscript on Mallock and his machine to me and allowed me generously to cite it. M. G. Croarken, 'The Mallock Machine,' unpublished manuscript, August 9, 1984. I am also obliged to Martin Campbell-Kelly who called my attention to the work of Mary Croarken. Cf. also M. G. Croarken, *Early Scientific Computing in Britain* (Oxford, 1990), p. 49 ff.
51. R. R. M. Mallock, 'An Electrical Calculating Machine,' *Proceedings of the Royal Society of London*, Series A, Vol. 140 (1933), 457–483, here 457.
52. As Wilkes reports, in 1937 the machine could solve a system of 10 equations with 10 unknowns. M. V. Wilkes, *Memoirs of a Computer Pioneer* (Cambridge, MA, London, 1985), p. 29.
53. Cited by M. Croarken, 'The Mallock Machine ...,' p. 8.
54. Croarken, 'The Mallock Machine ...,' pp. 7ff.
55. Cf. M. G. Croarken, 'The Emergence of Computing Science Research and Teaching at Cambridge, 1936–1949,' *IEEE Annals of the History of Computing*, Vol. 14, No. 4 (1992), 10–15. Croarken, 'The Mallock Machine,' op. cit., p. 3. The English pioneer of computers, Maurice V. Wilkes, reports in his memoirs how Mallock's machine was bought for the Mathematical Laboratory of the University. The lack of precision, arising from the losses in the transformers, was adjusted by a sophisticated controlling device, the so-called 'compensator' which Wilkes estimated, looking back

with some respect but also a bit relativized, 'As a piece of electronics, this was well ahead of its time.' Wilkes, *Memoirs*, op. cit. P. 29.

56. Cf. M. J. G. Cattermole, A. F. Wolfe, *Horace Darwin's shop. A history of The Cambridge Scientific Instrument Company 1878 to 1968*, (Bristol, Boston, 1987), pp. 136–139. R. R. M. Mallock, 'An Electrical Calculating Machine ....'
57. Anon., 'The Mallock Electrical Calculating Machine,' *Engineering* 137 (1934), 698–700.
58. V. Bush, 'Recent Progress in Analysing Machines ....'
59. Cf. A. C. Aitken, 'Mr. Mallock's Electrical Calculating Machine,' Letter to the editor, February 9, 1935, *Nature* 135 (1935), 235.
60. Croarken, 'Mallock Machine ...,' p. 12.
61. Croarken does not believe the the correctness of some reports which say that the machine had calculated determinants to degree 11 because Mallock nowhere mentioned how it was possible to detect that a determinant was zero.
62. At this time Alain Turing was working at NPL on his computer design and took trouble of its realization. On the other hand, the NPL still some years later gave an offer to the German company Schoppe & Faeser for the design and production of perhaps the biggest large-scale mechanical Differential Analyzer, which was ever built, and which was delivered in 1950. Obviously this machine should never be used. Cf. *Petzold, H.*, Rechnende Maschinen. Eine historische Untersuchung ihrer Herstellung und Anwendung vom Kaiserreich bis zur Bundesrepublik, Düsseldorf 1985, 76–81.
63. Wilkes, 29.

David Rhees, Kirk Jeffrey

Earl Bakken's Little White Box: The Complex Meanings of the First Transistorized Pacemaker

On a brilliant August morning in 1994, a large crowd of officers, employees, and stockholders of the medical device firm Medtronic, Inc., gathered outside the company headquarters in a northern suburb of Minneapolis, Minnesota, U.S.A., for an important ceremony. The occasion was the retirement from the board of directors of Earl E. Bakken, who in 1949 had co-founded the company and had helped make it into the world's leading manufacturer of cardiac pacemakers.[1] There were many speeches that morning, but the highlight of the ceremony was the unveiling of a full-sized statue of Bakken that stood facing the main door of the company headquarters building. Partially encircling the statue was a low stone wall engraved with the mission statement that Bakken had written in the early 1960s, which was still in force more than three decades later. While the words spelled out the values for which the company had become famous in the medical community, the visual symbol of Medtronic's corporate culture was held in the statue's outstretched right hand. There, forged in bronze, was a replica of the device that Bakken had invented during the winter of 1957–58—the world's first wearable transistorized cardiac pacemaker.[2]

The bronze replica in the statue's hand represents the crude prototype that Bakken made at the request of a renowned heart surgeon at the University of Minnesota. This prototype, housed in aluminum and containing only two transistors, had been intended for tests with dogs but was used on human patients within days of its invention. Soon afterward, Bakken and his employees introduced a more refined version of the transistor pacemaker in a black plastic shell; about ten of these went into clinical use at the University. Later in 1958, Medtronic began manufacturing a commercial version in white plastic—the '5800.' All three versions were essentially identical in circuitry and other interior features. The white production model was best known to doctors in the U.S. and abroad, but it is the earliest model that seems to hold the most meaning for Earl Bakken and others at his company.[3]

For Medtronic, this first pacemaker has come to embody the firm's creation myth. (We use *myth* in the sense of a story that gives meaning to the collective experience of a particular group.) During Medtronic's first decade, the company had led a precarious existence as a repair service for hospital electrical equipment and a regional distributor for a manufacturer of electrocardiographs. Medtronic also customized

Figure 1. Bronze statue of Earl E. Bakken holding a prototype of his transistorized pacemaker, at Medtronic headquarters near Minneapolis, Minnesota. Courtesy of Karen Larsen.

standard instruments and built new ones to order for laboratory and clinical researchers at the University of Minnesota and a few other sites in the upper Midwest. Bakken's transistorized pacemaker, made at Dr. C. Walton Lillehei's request to treat an unusual complication of open-heart surgery, seemed at first to be just another one of these special orders. This order, however, opened up what would prove to be an enormous new market in medical electronics, positioning Medtronic to become the world's leading manufacturer of invasive electrotherapeutic devices. In retrospect, Earl Bakken's invention of the transistorized wearable pacemaker in 1957–58 marked a new beginning for the company. For Bakken and many others at Medtronic, the pacemaker also symbolizes their shared belief in medical progress through technology. So it is not surprising that Medtronic would permanently enshrine the event in bronze.

Earl Bakken's retirement ceremony evokes an important theme of this paper, that people often impute powerful cultural meanings to made things. Numerous scientific and technological artifacts have achieved such iconic standing: one thinks of Lavoisier's chemical equipment at the Conservatoire des Arts et Métiers in Paris, the apparatus in the Galileo Room at the Deutsches Museum, the Newcomen and Watt steam engines at the Science Museum in London, and the John Bull locomotive at the Smithsonian Institution. However, we know relatively little about how such objects have acquired their cultural meaning. On one point there seems general agreement: various groups will understand the core significance of an object in somewhat different ways. In his recent study of scientific symbols and cultural meanings in American life, sociologist Christopher P. Toumey argues persuasively that symbols are 'polysemic,' meaning that 'they are capable of conveying multiple different meanings so that the same image means different things to different people.' Social groups may also attribute a variety of meanings to *artifacts*. It is this process of endowing technological objects with cultural meanings that most intrigued us as we explored the history of the Medtronic 5800 pacemaker.[4]

This essay asks how the construction of meaning occurred in the case of Earl Bakken's transistor pacemaker, not only within Medtronic's corporate culture but also among various other social groups that encountered the artifact. Other studies have elucidated the meanings of *generic* technologies to various social groups; Pinch and Bijker's study of the bicycle and Bettyann Holtzmann Kevles's recent history of medical imaging technologies come to mind.[5] We focus as much as possible on the meanings of a specific artifact, or rather its three physical embodiments. We also seek to include as broad a spectrum of relevant audiences as our sources permit: the inventor and his company, the surgeon and his university, other physicians who used these devices, some of the early patients, and the broader public.

Melvin Kranzberg pointed out in 1986 that the history of technology 'is a very human activity'; by implication, historians of technology also constitute a relevant social group.[6] Thus empowered, we suggest some meanings not apparent to the other audiences for Bakken's pacemaker. Business historians Louis Galambos and Jane Eliot Sewell have shown how basic scientific advances in bacteriology, virology, and genetic engineering have repeatedly sparked cycles of innovation that in some cases have lasted for decades.[7] Bakken's pacemaker, as one of the first successful applications of transistor technology to medical devices, helped launch a new innovation cycle in the field then called 'medical electronics.' The invention represented a transition from the previous generation of vacuum-tube, tabletop equipment connected to the electrical power system to a new generation of solid-state, battery-powered devices that could be attached to the human body and very soon fully implanted within it. This remarkably fruitful cycle of innovation, driven in part by physicians' repeated discoveries of new diseases in new patient populations, continues to the present day.[8]

The practice of cardiac pacing to treat bradycardias (disorders of heart rhythm in which the heart beats too slowly) eased itself into American medicine between the early 1950s and the mid-1970s. During these years, the first pacemaker implantations took place and cardiac pacing grew from an unusual and daring procedure reported in national newspapers and magazines to a reliable and routine treatment for many heart rhythm disorders. By the 1980s, the implanted cardiac pacemaker had become a commonplace medical device in the United States, Canada,

Figure 2. Earl Bakken with prototype of the first wearable pacemaker, 1997. Photograph by Thomas K. Perry.

western Europe, Israel, Australia, and most other developed countries, though implantation rates continue to vary greatly from country to country. No comprehensive registry of implantations exists, but we estimate that roughly two million people in the world carry pacemakers in their bodies today.

The expanding demand for cardiac pacemaker therapy has made possible a thriving industry with current worldwide sales of more than $2.5 billion. The Twin Cities of Minneapolis and St. Paul, Minnesota, constitute the leading center for the industry. Medtronic, the leader in implantable pacemaker sales from 1961 to the present, has given rise to more than thirty new medical technology companies staffed by former Medtronic employees. Two of these companies, St. Jude Medical and Cardiac Pacemakers, Inc., the latter a division of Guidant Corp., are Medtronic's chief competitors today in the world pacemaker industry.[9] Medtronic and the many 'baby Medtronics' have contributed to the emergence of the Twin Cities' reputation as 'Medical Alley,' a world-class center for research and manufacturing in medical technology. The Medtronic 5800 pacemaker thus led to big things a few years after its invention.

Because of its ubiquity, Americans and western Europeans are today perhaps inclined to take for granted the pacemaker and even the implantable defibrillator or ICD (an invention of the 1980s). They might be surprised to learn that in the entire history of medicine before 1957, there had never been a partly or completely implantable electrical device. In the development of cardiac pacing as a medical field, the invention of a transistorized external pulse generator in 1958 was an important step that few remember today. But to Earl Bakken, the people at his company, and many doctors and patients, the encounter with electrostimulation was immensely important; it remains vivid and significant to them four decades later.

Clinical Background: Open-heart Surgery at the University of Minnesota

The earliest attempts to pace the human heart date back to the late 1920s, but practical equipment had to await improved understanding of heart rhythm disturbances. Pacing did not acquire medical respectability until the postwar period, when an array of new and sophisticated biomedical electronic equipment came into use. Many of the new machines used technology developed from World War II military research efforts. Electrocardiographs, electroencephalographs, electromyographs, spirometers, defibrillators, ultrasound imaging, physiological integrators, and other electrical instruments and devices gradually diffused into everyday use in hospitals, clinics, and research laboratories. The field of medical electronics came into being, with its own journals, texts, and training programs. By the early 1960s, it was generally taken for granted that the well-trained physician in cardiac acute care would make extensive use of electronic equipment: not only pacemakers but defibrillators, cardioverters, and heart monitors. Cardiologists Desmond Julian in Edinburgh and Hughes Day in Kansas City independently organized special hospital coronary care units staffed by trained teams of doctors, nurses, and technicians and equipped with the latest equipment to monitor heart attack victims and resuscitate them if their hearts slipped into dangerous arrhythmias. By 1967 there were 350 such units in the United States.[10]

In 1952, as medical scientists and clinicians were beginning to embrace this new set of technological marvels, a cardiologist in Boston named Paul M. Zoll invented a tabletop pacemaker with chest electrodes and successfully used it to resuscitate people from standstill of the heart in the hospital. The proximate origins of cardiac pacing as a medical field and an industry lie in Zoll's work.[11] The Medtronic 5800, however, was invented a few years later to deal with an unexpected side-effect of open-heart surgery. At the University of Minnesota, under the driven leadership of Dr. Owen Wangensteen, the surgical faculty and their trainees had transformed the Department of Surgery into an invention factory of international significance. After World War II, the university trustees and private groups had contributed $1.5 million to build a special hospital dedicated to the treatment of acute heart disease. This facility, the first of its kind in the United States, opened in 1951. As chief of surgery, Wangensteen insisted that every member of the surgical faculty pursue some program of research and that every surgical resident contribute as an apprentice investigator. Generous funding for surgical research was available throughout the 1950s from private sources (the American Heart Association) and public (the National Heart Institute within the National Institutes of Health).[12]

Surgeons at Minnesota developed or improved numerous techniques for entering the living human heart and correcting congenital defects; these efforts culminated in 1954–55 with the first successful surgery for total correction of tetralogy of Fallot, a set of four defects that weakened and stunted children (these were the famous 'blue babies'), then usually killed them in adolescence.[13] In order to perform the lengthy and complex corrective operation, it was necessary to stop the heart from beating; this meant devising some temporary means to provide a circulation of oxygenated blood to the patient. The surgical research program at Minnesota developed not one but three new technologies that enabled surgeons to work for many minutes within the heart.[14] The third of these, a true heart-lung machine known as a bubble oxygenator, was ready for use with human beings in May 1955.[15]

The leading heart surgeon at Minnesota, C. Walton Lillehei, was an intense, charismatic figure who had attained international fame by the mid-1950s.[16] By 1957, as Leonard Wilson vividly recounts in *Medical Revolution in Minnesota*, Lillehei and his assistants had carried out 305 open-heart operations on young adults and children. However, even when he successfully repaired a defect, about one patient in ten developed complete heart block as a consequence of the surgery itself. In heart block, the electrical impulse that begins high in the right atrium of the heart fails to reach the main pumping chambers, the ventricles. Deprived of their normal signal, the ventricles may beat slowly on their own, but their rate gradually falls. Eventually the heart fails to provide an adequate circulation of oxygenated blood to the brain and the body and the patient dies. The surgeons concluded that they were occasionally causing damage to the heart's conduction pathways while carrying out their surgical repairs. But they believed that if they could keep patients alive for two to three weeks with artificial support for the heartbeat, patients' conduction systems would heal and normally conducted beats would resume. Zoll's external pacemaker was clearly inappropriate because it delivered impulses at 50 to 150 volts through electrodes strapped to the patient's chest. For children, 'it was way too traumatic,' Earl Bakken recalled. 'They'd be bouncing on the table each time the pacemaker fired.'[17]

The Myocardial Wire

As one of Lillehei's surgical trainees later pointed out, open-heart surgery was all new and 'the learning process was one of trial and error.' When external pacing proved a disappointment, Lillehei's group tried several drugs that stimulated the heart to beat. Now his success rate rose: out of seventeen children in heart block and treated with drugs, nine survived long enough to revert to normal heart rhythm, five remained in heart block but survived, and five died. 'For our purposes, that was a nice improvement,' Lillehei has recalled; but it 'was obviously not satisfactory.'[18]

Around the summer of 1956, Lillehei and his associates decided to try pacing the children's hearts through an electrode that would actually touch the surface of the heart. By delivering the electrical stimuli directly to the excitable tissue rather than firing them through the body from outside, this arrangement might capture the heartbeat at a much lower voltage. Vincent L. Gott, one of Lillehei's trainees, borrowed a laboratory physiological stimulator and took it to the dog lab. Using a standard surgical technique he created heart block in a dog. He inserted a wire into the heart wall (the myocardium), connected it to the external stimulator, and found that 'it picked the rate right up.'[19]

Lillehei wondered whether the heart tissue would develop a rejection reaction to the metal electrode or the wire cause a serious lesion as it moved around with the beating heart, but he went ahead anyway because 'we were desperate.' On January 30, 1957, he implanted a multistranded, braided stainless steel wire in a Teflon sleeve into the ventricular myocardium in a three-year-old girl when she developed symptoms of heart block during open-heart surgery. He brought the other end out through the surgical wound, attached it to an external stimulator, and buried an indifferent electrode under the child's skin to complete the circuit. The little girl survived.[20] Lillehei soon came to rely on the myocardial wire whenever a patient showed signs of heart block at the end of an open-heart operation. By delivering the electrical pulses directly to the excitable tissue of the heart, this arrangement permitted effective pacing at around 1.5 volts. Days later, when the heart's own conduction fibers had healed, the surgeon could tug gently on the wires and pull them out of the child's body.

But maintaining a heartbeat through a myocardial wire had problems of its own, the most important being that the pulse generator was still a bulky tabletop stimulator plugged into an electrical socket. Lillehei wished to get the children out of bed and was concerned because their hearts were potentially exposed to power failures and surges in the hospital electrical system and short circuits in the pulse generator. On 31 October 1957, an equipment failure at a large Twin Cities power plant caused an outage in Minneapolis that lasted more than two hours.[21] The University hospitals had auxiliary power, but Lillehei viewed the event as a warning. Even before the power failure, he had asked a graduate student in physics at the university to make a pulse generator powered by a battery. When the student failed to produce a device after repeated prodding, the surgeon turned to Earl Bakken, the electrical engineer who repaired and calibrated electronic equipment for the Department of Surgery.[22]

The Inventor: Earl Bakken and Medtronic

Born in 1924 and raised in Minneapolis, Earl Bakken had interrupted his college years to serve in World War II, then completed his B.S. in electrical engineering at the University of Minnesota after the war. He began taking graduate courses, and through his wife, a student in a program to train medical technicians, soon found himself being asked to repair equipment at Northwestern Hospital and at the University Hospital. Few hospitals at that time employed trained electronic technicians; a skilled repairman who did not get sick at the sight of blood would have found it easy to get work. Bakken recognized a business opportunity opening up. With his wife's sister's husband, Palmer J. Hermundslie, who had been in the lumber business, Bakken founded Medtronic in 1949; the name was derived from combining 'medical' and 'electronics.'[23]

Bakken found that researchers at the University of Minnesota Medical School and the nearby campus of the College of Agriculture. Investigators often 'wanted special attachments or special amplifiers' added to some of the standard recording and measuring equipment. 'So we began to manufacture special components to go with the recording equipment. And that led us into just doing specials of many kinds. For the farm campus, for the X-ray department, we developed many, many instruments: animal respirators, semen impedance meters for the farm campus, just a whole spectrum of devices.' Usually Bakken's company would sell just a few of these items. Medtronic by 1957 had four or five employees and was located in a remodeled three-car garage that stood behind Hermundslie's parents' house in north Minneapolis. 'It was a desperate struggle all the time to meet the next payroll,' Bakken has said. 'We would make these specials for people, and they'd always end up costing us much more to make than we had bid. Everything we did, we lost money on… . We just seemed to go further and further downhill and borrowed more and more money.'[24]

When Bakken accepted Lillehei's assignment in December 1957, it seemed to the engineer just another special order. His first thought was to set up a cart that would hold equipment to run a conventional AC-powered pacemaker: a six-volt DC automobile battery, a battery charger, and an inverter. But then he realized that he could build a heart stimulator using transistors, a few other components, and a small battery. He borrowed a circuit design for a metronome that had appeared a the year before in an electronics magazine for hobbyists.[25] He used a 'powerful miniature [9-volt] mercury battery,' housed the assemblage in an aluminum circuit box, and provided an on-off switch and control knobs for stimulus rate and amplitude.[26] Bakken assumed that the surgeons would test the device by pacing laboratory dogs—his company had no animal testing facility. They did 'a few dogs,' then Lillehei put the pacemaker into clinical use. When Bakken next visited the university, he was surprised to find that his crude prototype was managing the heartbeat of a child recovering from open-heart surgery.[27]

Readying the Pacemaker for the Market

As a regional distributor for the Sanborn company, Earl Bakken regularly visited surgical departments throughout the upper Midwest and recognized that there might be a market for the battery-powered pulse generator. In the spring of 1958, he and others at Medtronic did what they could to redesign the device as an attractive product for hospitals that were setting up open-heart surgery programs. The first commercial model had recessed knobs to prevent the children from changing their own heartrates. It also sported two little handles so that straps could secure the device to a patient's chest; Bakken had borrowed the handles from an old ECG machine to create a device that was not only portable but *wearable*.[28] The pulse generator also had a little neon light that blinked red with each stimulus—a feature that we shall discuss further below.

The housing of the 5800 'was a kind of carvable Bakelite paneling that was available at the time. It was layered, white inside, black on the back, and then it was carved' to bring out the white lettering. Sometime in the

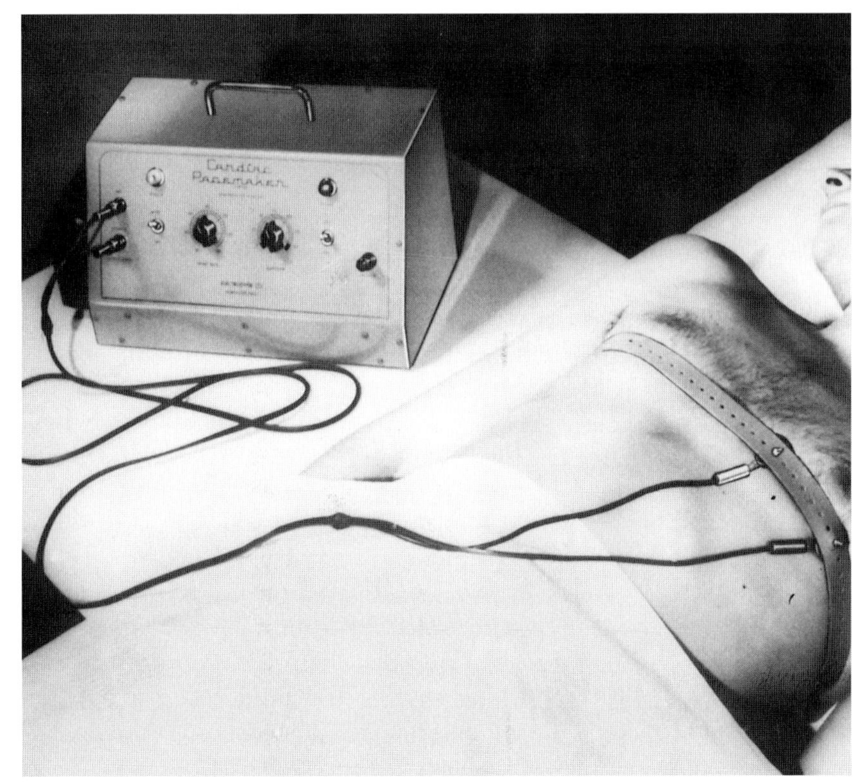

Figure 3. The first commercial, external, AC-powered cardiac pacemaker, ca. 1953–54, invented by Dr. Paul Zoll and manufactured by the Electrodyne Co. This unit delivered shocks that were too powerful for children. Courtesy of Medtronic, Inc.

spring of 1958, Bakken ordered an important design change in the product: by reversing the Bakelite housing, he changed its color from black to white. 'It appeared to me that white was more appropriate with the whiteness and cleanliness of a hospital,' Bakken later commented. He also joked that 'the black didn't work out all that well because you could always see it; and if it was pinned to the laundry or to the sheet of a child's bed, before they threw the laundry in the chute to have it cleaned, the pacemaker always got removed. So when you make it white, then it isn't so visible—so they throw them away and have to buy more—so that was an advantage.'[29]

The surgical program at Minnesota in effect marketed the 5800: scores if not hundreds of surgeons visited Minneapolis to observe Lillehei's team at work, and each year surgical residents who had trained under him fanned out to positions at leading hospitals and medical schools. A few hospitals in the United States and western Europe that were setting up programs in open-heart surgery sent in orders for the pacemaker, which was known as the 5800 'because we made it in 1958.' In 1960, Lillehei and others at the University, together with Bakken, published a paper in the *Journal of the American Medical Association* that described the device and discussed its clinical uses. Two illustrations showed the 5800 and the Medtronic name up close; one of them pointed out nine important features of the pacemaker such as its handles, neon flasher, control knobs that 'cannot be accidentally changed,' and white case that 'allow[s] for damp scrubbing with alcohol.' No ad agency could have prepared a more effective advertisement for the invention.[30]

Medtronic probably sold only a few hundred of the 5800 pulse generators between 1958 and 1964. The reason was that the device was not implanted but could be re-used to assist many patients over a year or two, whereas implantable pacers (introduced late in 1960) were used only once. Heart surgeons sometimes resisted postsurgical cardiac pacing; Lillehei has suggested that some were probably reluctant to admit that they were creating heart block when putting stitches into the ventricular septum, and many clearly didn't like the idea of having to manage the myocardial wire and the external pulse generator, even for a few days. Bakken remembers sitting in the Medtronic booth at a medical convention and noticing that heart doctors would walk on the opposite side of the aisle to avoid having to talk to him. He thinks that they didn't want to reveal their ignorance of electronics and their dependence on an engineer to enlighten them. In later years, doctors swallowed their pride: today company technicians attend and assist at most pacemaker implantations in the U.S. Even in 1958, it seems, the pacemaker with its two transistors heralded an era when physicians would come to rely on various kinds of experts to invent and manage the technologies of cardiovascular medicine.[31]

The 5800, the Heart, and Medical Progress

The Medtronic 5800 contributed to the later development of cardiac pacing, but it is best understood in the context of its own time. Ten years earlier, the idea of connecting a pulse generator to a wire sewn into the wall of the heart itself would have been unacceptable to physicians anywhere in the world. Permitting a person so unfortunate as to be dependent on such a device to get out of bed, walk the hospital corridors, and perhaps even go home would have been inconceivable. But by the late 1950s, surgeons had learned that the heart is quite sturdy; they had developed a sense of confidence about going through the pericardium, working around the exterior of the heart, and even cutting into the heart itself.

Many organs of the body are necessary to sustain life, but European and American cultures have imbued the heart with special cultural meanings. For centuries, we have imagined the heart as the seat of the human emotions, particularly humane feelings toward others, and of the soul itself. This mystique long antedates Harvey's discovery that the heart pumps blood throughout the body. Even today the English language retains scores of metaphorical expressions of the ancient belief that the heart is the mysterious center of our emotional natures, our very identities. Pre-scientific and romantic notions about the heart have not disappeared but linger in modern culture at the end of the twentieth century. For example, men and women scheduled for heart transplants often ask if their new hearts will cause changes in their personalities, particularly if the donor was of the other sex or a different ethnic origin.[32]

Whether they continued to believe in traditional verities or not, Americans readily accepted open-heart surgery, heart transplants, and machine substitutes for the heart. Indeed, perhaps their belief that the heart was a sacred part of the body helps explain why many Americans developed an attitude approaching reverence for the God-like physicians who cut it open. But at the same time, advances in heart care since 1945 were founded on a mechanistic understanding of the heart and a more aggressive, manipulative approach toward it. Behind them all lies the belief that the heart is nothing but a pump, a machine within the body. This shift in medical thinking had begun in the age of Descartes and Harvey; it had made possible the laboratory studies of the heartbeat as an electrical phenomenon and the invention of the electrocardiograph at the end of the nineteenth century.[33] Heart surgery and the other new treatments of the postwar period did not inspire but rather built upon this understanding of the heart as a piece of machinery. Thinking of the heart as a pump leads on to the possibility that we can open it up and repair it when it breaks down, perhaps by replacing worn-out parts of the machinery. Despite its apparent simplicity, Earl Bakken's transistorized pacemaker of 1958 embodied a set of attitudes about the heart that still inform the cardiovascular technology industry and the practice of cardiovascular medicine.[34]

In a rapidly changing field like cardiac pacing that has brought wealth and renown to many and 'full life' to many more, all participants and most observers will understandably tend to view the history of the field as a case study in technological progress.[35] From this premise, it is easy to move on to the belief that each technological innovation has represented an inevitable step toward ever more advanced, sophisticated, and subtle cardiac pacing devices. Our own belief is that impersonal explanatory concepts such as 'medical progress' or 'the advance of technology' offer little insight into why some inventions won general acceptance while others were rejected and soon forgotten; and by invoking powerful 'forces' apparently unrelated to human choice, such explanations imply that the role of human beings has consisted of conforming themselves to the inevitable.[36] We believe that people working in specific cultural situations, after all, invented pacemakers, and that each new feature or new device had to satisfy other people—physicians, above all—whose reasons for accepting or rejecting new medical devices often went beyond narrowly technical factors.[37]

The progressive view of steady and inevitable technological progress in cardiac pacing also takes for granted that inventors and the social groups for whom they worked always and consistently shared a common goal, a stable vision of what the pacemaker was for and how it ought to develop. But this has clearly not been the case: research into heart rhythm problems has repeatedly led to the framing of new disorders for which some form of cardiac pacing seemed the appropriate treatment. At first these disorders all involved unduly slow heart rates; more recently, some pacemakers can detect and halt heart rates that are too fast (tachycardias). Again and again, heart specialists have revised their ideas about what pacing was for.

Consider the key social group in 1958, the cardiothoracic surgeons who pioneered open-heart surgery in children. While many were cautious about the new device and some perhaps avoided the Medtronic booth at medical conventions, most eventually accepted the need for postsurgical pacing sooner or later. The technical advantages of the transistorized pacemaker for managing the heartbeats of children recovering from open-heart surgery seemed obvious compared with alternative treatments or no treatment at all. Lillehei spoke of his first use of the myocardial wire as 'a revelation'—that a simple piece of wire 'could drive the ventricles very accurately... . It was great!' Others too spoke of 'driving' the heart. By reducing the daunting complexity and uncertainty that surgeons faced, it helped free the surgeon to focus on other aspects of open-heart procedures and the care of his patients. But the 5800 was appealing for other reasons as well. These men (they were all men) enjoyed national acclaim at that time; they had the image of being quintessentially modern and high-tech, and the Medtronic external pacemaker had *modernity* written all over it. It was small, white, transistorized, a technologically satisfying solution; and it was associated with a radically new practice, management of a physiological function over a period of time through electrostimulation via a wire left in the heart.[38]

The context within which the 5800 constituted a step in medical 'progress' was actually more complex than the foregoing suggests: at the very time that the device came into use, a debate was underway among clinical researchers over whether the future of cardiac pacing lay within the hospital for short-term stimulation or whether, on the other hand, patients might someday be able to leave the hospital and lead 'normal' lives while relying on pacemakers to manage their heartbeats. Invented for short-term pacing in the hospital, the 5800 was soon redefined as a device for long-term heart stimulation outside the hospital. By finding new uses for the device, its users redefined cardiac pacing itself.[39]

The 5800 as an Example of Innovation in Medical Devices

The invention and early use of Bakken's external pacemaker could serve as a textbook case illustrating how medical innovation worked in the (American) real world between World War II and the imposition of federal regulatory oversight for life-sustaining medical technologies in 1976.[40] First, the pacemaker like most new medical devices embodied *the transfer of technologies* developed in different realms for entirely different artifacts such as metronomes and flashlights. The blocking oscillator circuit that Earl Bakken borrowed from *Popular Electronics* had actually been invented at the MIT Radiation Laboratory during World War II. Later implantable pacemakers used tiny mercury-zinc battery cells invented during the war for Army field telephones (walkie-talkies). The implantables also used Scotchcraft epoxy from Minnesota Mining & Manufacturing (3M) and Silastic silicone rubber from Dow Corning to shield electronic components from the harsh environment within the human body. In a broad sense, the pacemaker technology of the 1950s and 1960s was a civilian spin-off from the R&D of the wartime and Cold War eras. As Bakken put it, 'it was kind of an interesting point in history, a joining of several technologies.'[41]

Second, Annetine Gelijns and Nathan Rosenberg have pointed out that with medical devices, the distinction between a stage of research and development (R&D) and a stage of adoption often breaks down in practice: 'It is a serious misperception,' they write, 'to think that all important uncertainties have been ironed out by the time a new technology has finally been introduced into clinical practice. In fact, much uncertainty associated with a new technology can be resolved only after extensive use in practice.'[42] Bakken's transistor pulse generator made *a nearly overnight transition from bench testing to clinical use*; this set a pattern in the cardiac pacing industry. For the next decade at least, it would be a common practice to put new devices, including fully implantable ones, into clinical use and then iron out the imperfections based on the accumulation of clinical experience. This practice developed because most of the early patients were close to death, no other treatment existed, and American medical culture rewarded physicians for using bold new strategies.[43]

A third pattern of broader application was apparent in Minneapolis during the late 1950s. This was *mutual dependence between device manufacturing firms and their clients.* The clients were not the patients kept alive on products like pacemakers but the physicians who decided whether to embrace a new technology such as cardiac pacing, and then which specific pacemaker brands and models to use. Cooperation between innovative physicians and device manufacturers was founded on mutual dependence; it was present in embryonic form in the relationship between Earl Bakken and Walt Lillehei in 1957–58.

Lillehei was a famous surgeon, Bakken an unknown who tinkered with electronic equipment and repaired things. However, surgeons could not invent electronic medical devices themselves, so they had to rely on the people who could. Bakken's company, Medtronic, stood out from a myriad of small electronics repair shops because of its association with one of the most famous surgical establishments in the world.[44] Bakken had begun attending surgical procedures at the university even before 1957 and had a personal locker in the surgeons' locker room. He carefully nurtured the friendships he formed at the university. He has observed that 'many of the residents, interns, that were working for Lillehei at the time, went on to become heads of surgery around the world.' He 'kept in good touch with physicians around the world; but it kept me traveling at a great rate, a good deal of the time.'

For the next decade, Medtronic's reputation and sales—its very survival—would depend heavily on public associations between the company and distinguished surgeons and engineers who served as consultants, and on Bakken's personal acquaintance with leading surgeons. These connections gave Medtronic a kind of instant credibility in the world of cardiovascular medicine. From the 1970s on, Medtronic and its competitors institutionalized their contacts with leading physicians by offering financial support for clinical research, inviting physicians to participate in the clinical testing of new devices, and forming physician advisory committees to recommend directions for future product development. Today several leading physician-researchers hold management positions at Medtronic and four physicians are members of the board of directors.[45]

Finding New Uses for the 5800

A fourth pattern of interest, *the search for new uses for the invention,* leads us into the later history of the Medtronic 5800. About 0.8 percent of newborns have some kind of serious congenital malformation of the heart; in the 1950s, this amounted to thousands of children—but only a fraction

Figure 4. Earl E. Bakken (shown) and Palmer J. Hermundslie founded Medtronic in 1949 in a three-stall garage in Minneapolis. Courtesy of Medtronic, Inc.

had surgical corrections and only a fraction of these developed heart block.[46] What happened next was a broadening of the list of conditions for which the pacemaker would prove useful. Several research groups had been exploring the idea of long-term pacing since about 1956; in September 1958, they debated its feasibility at a one-day conference in New York City that Earl Bakken attended.[47] It was known that elderly men and women sometimes developed *chronic* complete heart block (as opposed to *acute* postsurgical block), though heart specialists were not certain just why this happened or how many cases there might be. A few investigators believed that many patients died from complete heart block before ever seeing a physician competent to diagnose the condition. As one wrote, 'the disorder with its ominous prognosis had taken its toll well before the patient was referred to the specialist. Since no effective therapy was available, the level of diagnostic suspicion was low.'[48] In a sense, the pacemaker and complete heart block were made for each other. Had there been no prospect of 'effective therapy,' interest in the disease and 'diagnostic suspicion' might have remained low.

In certain respects, the new pacing system—Lillehei's myocardial wire and Bakken's external pulse generator—still resembled Paul Zoll's original version of pacing from 1952. In both external and myocardial pacing, the patient was assumed to be gravely ill, confined to the hospital, and pacemaker-dependent. Both systems ministered to acute crises, coronary standstill or postsurgical heart block. In both, the pacemaker was defined as a piece of hospital equipment; its transformation into a more or less permanent addition to the patient's own body was still a few years away. But the Bakken pulse generator opened up new possibilities for cardiac pacing in ways that Zoll's original invention never could have done.

The existence of a plausible technology in the transistorized external pulse generator intensified the search to find uses for pacing.[49] One person who understood quite early that it might be possible to treat chronic complete heart block with the pacemaker was Norman Roth, a young engineer who had joined Medtronic's tiny work force in 1958. Roth also realized that the main impediment to long-term pacing was Lillehei's myocardial wire. It was 'just a stainless steel suture wire. And ... either the wire would break, or there was a tendency for fibrotic tissue to build up around the wire, and the resistance go up, and they would get to a point where you couldn't drive [the heart].' For either sort of failure, the only remedy would be to re-open the chest, remove the malfunctioning wire, and implant another. 'It was not a happy thought to have to reimplant an electrode'—particularly if the replacement electrode might fail within weeks.[50] Shortly after Bakken invented his transistorized pulse generator, Roth designed a new lead intended to permit pacing over many months. He paid particular attention to the end of the lead that would be in contact with the heart and designed a rectangular Silastic 'platform' measuring 1.5 by 2.5 cm, from which protruded two stainless steel pins, the anode and cathode of a bipolar electrode. Roth believed

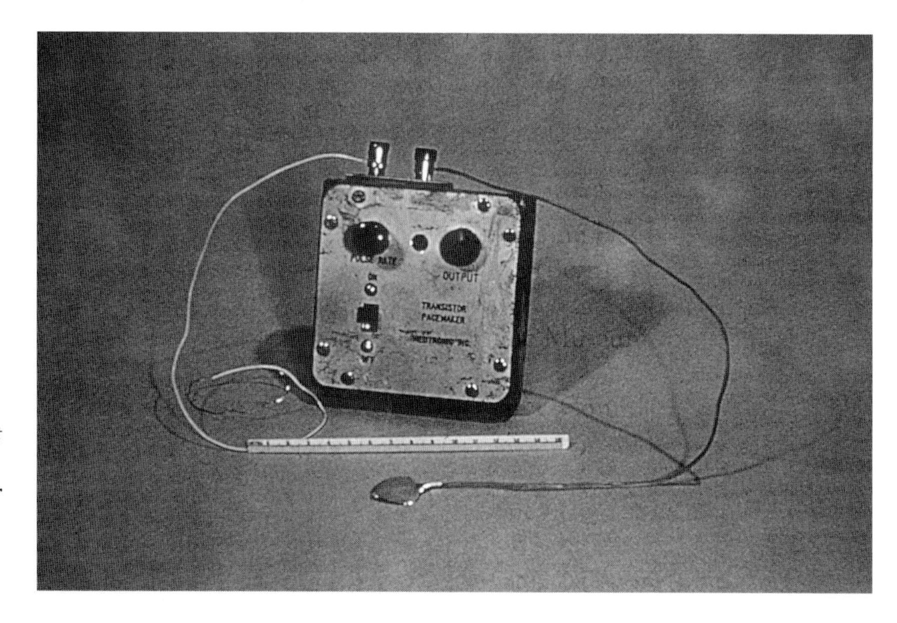

Figure 5. Earl Bakken's prototype of the first transistorized, wearable cardiac pacemaker. The wire was sewn to the heart and the flat indifferent electrode was placed under the skin.
Courtesy of Medtronic, Inc.

that when stitched down on the surface of the heart, the platform and bipolar configuration would provide a more stable interface between the pacemaker and the heart.[51]

When he showed a prototype of the lead to surgeons at the University of Minnesota in the autumn of 1958, they rejected it on the ground that 'we're only interested in something you can take out [of the patient's body]. And you can't take that out.' They were correct: it was impossible to remove one of Roth's electrodes simply by pulling on it. Roth next contacted a thoracic surgeon in St. Paul, Minnesota, Samuel W. Hunter, who had done a surgical residency under Lillehei in 1956–57. Hunter had the same reaction: 'I said, "You can't get it out! What are you going to do—leave it in there?"' But Hunter had a small animal research lab and was 'more or less casting around for things to do.' The same day that he met Roth, Hunter installed the platform electrode on the heart of a dog. When they connected it to an external pulse generator, it proved able to overdrive the dog's natural heartrate. Impressed, Hunter agreed to work with Roth on the animal studies. Together the surgeon and the engineer tested the new electrode in dogs by surgically creating heart block, implanting the electrode, and driving the dogs' hearts with a Medtronic 5800.[52]

A 72-year-old man in severe, chronic heart block was referred to Hunter in April 1959: the patient's ventricular rate varied between 16 and 36 contractions per minute and he was having dozens of episodes of coronary standstill a day. Roth prepared one of the experimental electrodes and Hunter implanted it. This patient, Warren Mauston of St. Paul, lived until October 1966 with a bipolar electrode and a Medtronic external pulse generator. He survived surgery for colon cancer, a severe attack of pneumonia, and an automobile accident. He always declined an implanted pulse generator, confiding to Hunter that his grandmother had refused indoor plumbing for her farmhouse because 'some things just don't belong inside.' Because of this choice, Mauston always had an open wound through which the pacing lead protruded. Hunter gave him antibiotics for a time but stopped when he grew concerned that this would encourage the growth of drug-resistant bacteria. Hunter or an assistant thereafter stopped at Mauston's house twice a week to check for infections and clean the site in Mauston's chest where the lead entered his body. About once a month, Mauston's generator would get a new battery: Norm Roth had 'modified the [generator] to have a large capacitor in it so that you could take the battery out and you'd still get eight or ten acceptable pulses—give you plenty of time to take the old one out and put the new one in.' The Hunter-Roth electrode served Mauston for four years, long past the time when Medtronic had withdrawn it as a commercial product. In 1963 Hunter abandoned the original lead and gave Mauston a transvenous lead that delivered the electrical stimulus within the right ventricle and emerged from the exter-

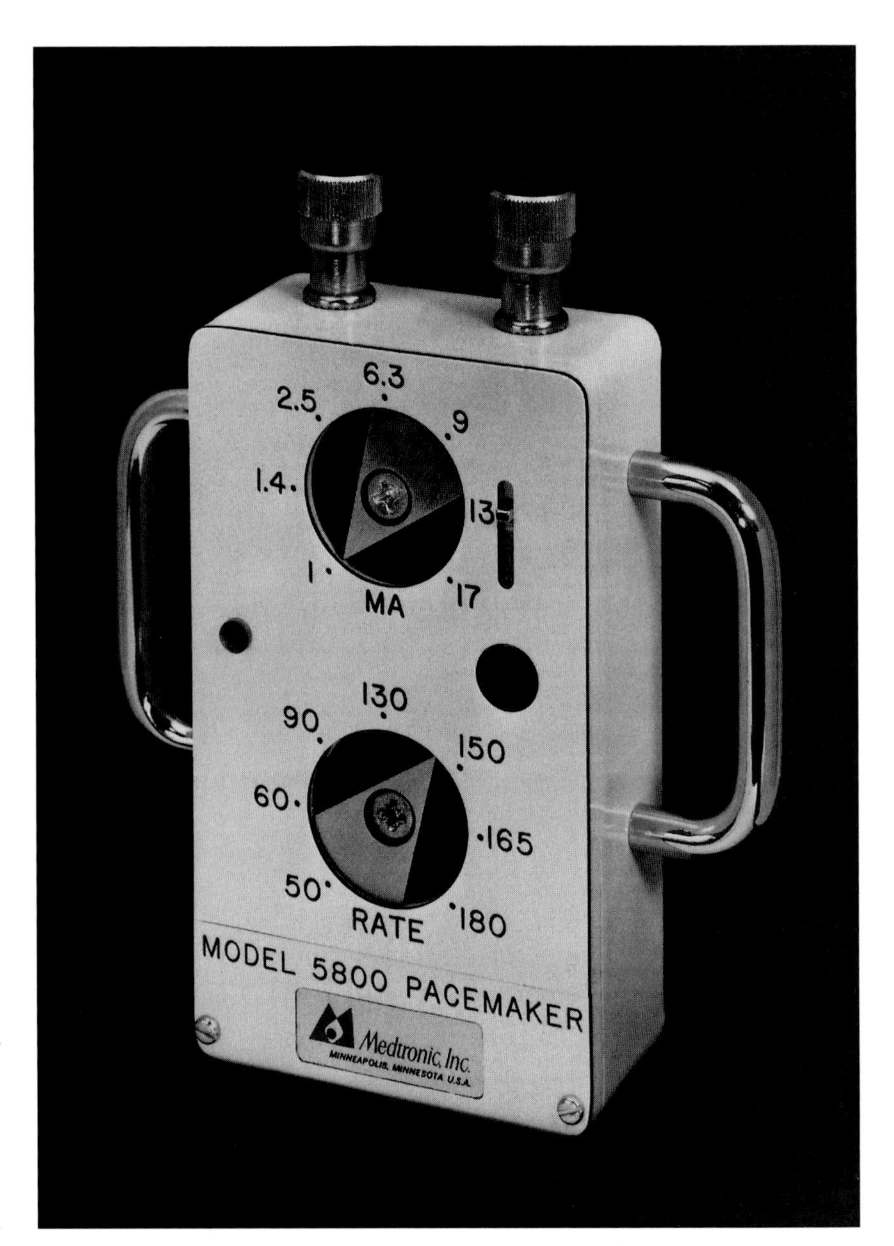

Figure 6. The Medtronic 5800 was the production model of Bakken's wearable unit. The white color was deemed more suitable for medical use. Note the recessed controls to prevent accidental adjustment. Courtesy of Medtronic, Inc.

nal jugular vein at the base of the neck. He continued to rely on an external Medtronic 5800 pulse generator. Mauston remained active until the very end. On the day before his death, he drove 60 miles to Lake City, Minnesota to visit his son and watch a World Series game on television.[53]

As they tried the Medtronic 5800 for postsurgical heart block or used it with elderly patients like Mauston, surgeons began to accumulate a body of practical experience in cardiac pacing. When they had managed a few cases, the early pacemaker doctors would present the results at medical conventions or in print, thus beginning the process of converting their experience into more formal doctrine. Roth traveled extensively to introduce the 5800 and the platform electrode at hospitals; Hunter gave a paper on the Mauston case at a national heart meeting. About eighteen months after he operated on Mauston, a surgeon on the West Coast called Hunter. 'He started to rail on me and used some very uncomplimentary words. He said, "This is the most jackass pacing equipment. I couldn't get the lead to fit into the external pacemaker… . This thing just is designed so poorly I can't believe it."' Hunter eventually realized that his caller 'had put the thing in backwards,' forcing the two spikes of the electrode into the sockets atop the external pulse generator and stitching the other end down on his patient's heart. 'He put it on backwards. And the funny part of it is, it worked.' Hunter added, 'I immediately wrote a paper with drawings [and] descriptions, showing how to put it on.'[54]

The case of Warren Mauston, along with a few others from the United States and England, had demonstrated that long-term cardiac pacing was possible. Until superseded by implanted pulse generators two or three years later, the Medtronic 5800 paced the hearts of at least 100 older men and women in chronic complete heart block.[55] More than that, the 5800 encouraged physicians to redefine chronic heart block, essentially to reconceive the disease. Physiologists had understood what happens in heart block since the early twentieth century, but their research had attracted little attention from clinicians because it yielded no practical clues as to how to treat the condition. Once Bakken, Roth, and Hunter had pioneered a plausible treatment, research into heart block picked up again and clinicians acquainted themselves with the symptoms of the disorder. This intensification of professional attention led in turn to the framing of still other diseases of the heartbeat for which a pacemaker seemed the appropriate treatment. The most recent set of formal guidelines for pacemaker implantation lists dozens of rhythm disorders, many of which had not been carefully studied or even noticed until after the invention of cardiac pacing.[56]

The 5800 and the Patient: a Technology of Reassurance

The physicians who implanted pacemakers were the true 'consumers' of the technology; their patients, for the most part, had no way to participate meaningfully in shaping the development of pacing. But in the early days, when pacing was very much a revolutionary and untried therapy, the patients did play an important role. The children and elderly men and women made clear to physicians, by their behavior, that they wanted to look and feel like healthy people even though their hearts refused to beat on schedule. As surgeon Seymour Furman has pointed out, 'a surprisingly large number did survive and even prospered,' sustained by the primitive pacing technology of the day and by their own determination. By

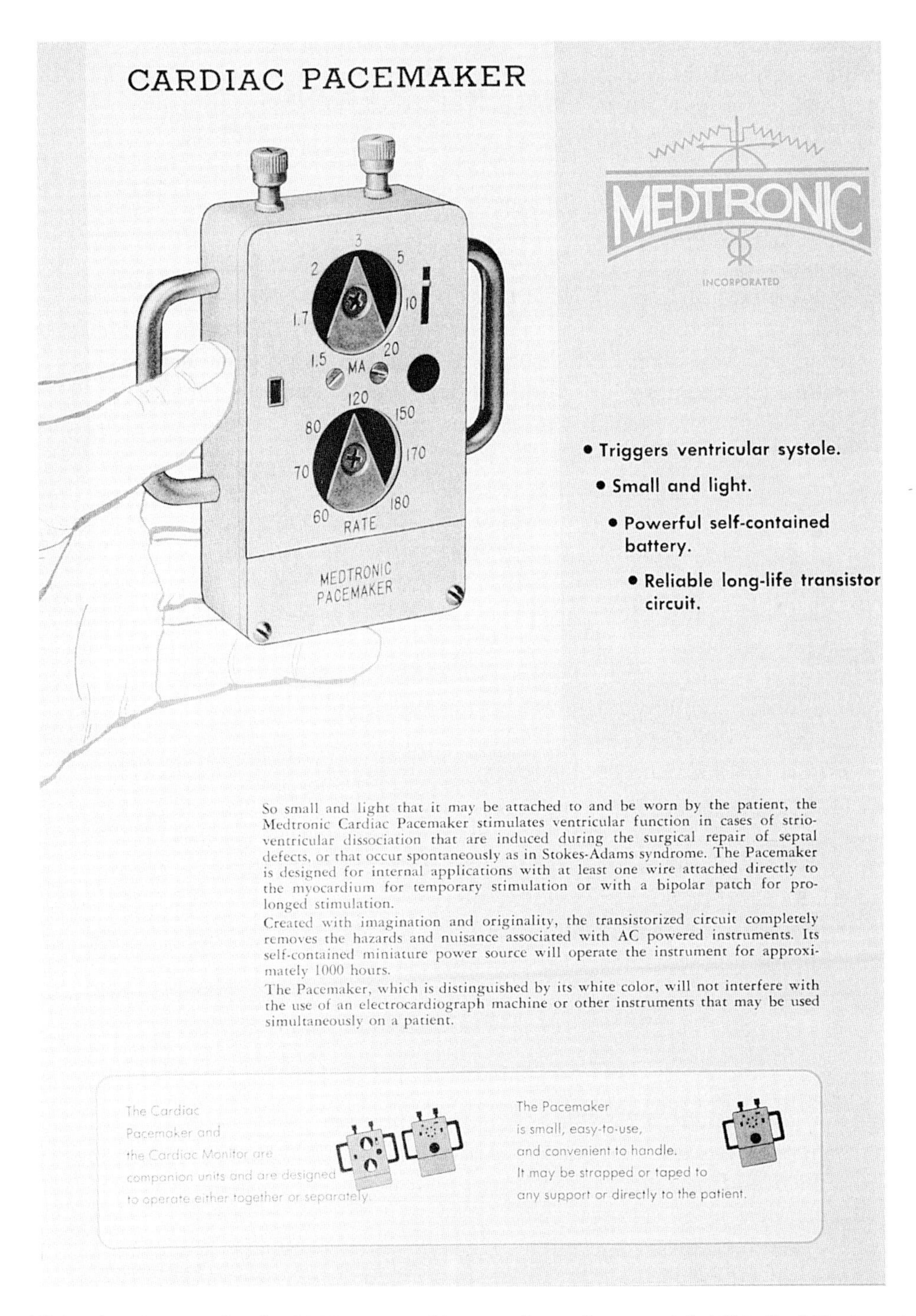

Figure 7. This advertisement for the 5800 appeared in a trade catalogue entitled 'Medical Engineering: Products and Services of Medtronic,' ca. 1960. The illustration was reprinted in C. Walton Lillehei et. al., 'Transistor Pacemaker for Treatment of Complete Atrioventricular Dissociation,' Journal of the American Medical Association *172 (30 April 1960): 2006–2010. Courtesy of Bakken Library and Museum.*

accepting pacemaker therapy (or, in a few cases, declining it and accepting the prospect of early death) and by trying to restore elements of normal life for themselves, the early patients helped the doctors and engineers to grasp the human significance of *permanently* pacing the heart.[57]

By the postwar period, the ordinary American was no longer in a position to comprehend the technological aspects of many important inventions from microelectronics to atomic fission because he or she lacked the necessary knowledge of the underlying scientific principles. But the citizen could in some cases assess a machine or artifact on its outward appearance and, of course, its effects. The Medtronic transistorized pulse generator received a surprising amount of attention in the national press, at least until fully implanted pacemakers were announced in mid-1960. Perhaps one reason is that, after all, it was external—you could see it and photograph it. Its workings were also relatively easy to comprehend. One feature above all intrigued people who observed the 5800: its blinking red light.[58] According to Earl Bakken, the flashing light reassured both physician and patient that the device was really stimulating the heart. 'We went to putting [in] a screw switch ... so that people could turn the light off because it would double the length of the battery. Nobody would do it.'[59]

To the patient, the patient's family, the national press, and the general public, the Medtronic 5800 apparently seemed a reassuring technology. Bakken's own pastor told the inventor that God did not mean for people to have machines in their bodies, and some people wrote letters to the editor along the same lines. But we doubt that this was the majority opinion at that time. As we noted earlier, open-heart surgery, the heart-lung machine, and the pacemaker originally constituted a technological package that addressed the needs of a population of children, the 'blue babies' who were so much in the news in the 1940s and 1950s. The plight of these children, like the similar plight of polio victims, served to focus widely shared American anxieties about the risks that innocents faced as well as hopes that medical research and technology might find the cure for these disabilities.[60]

The advent of open-heart surgery provided hope for children hanging on to life with congenitally malformed hearts. But as with polio victims in their iron lungs, it must have seemed to some families and other onlookers that the technology of treatment was nearly as dismaying as the condition itself. In the case of open-heart surgery, the child had to undergo hypothermia (under anesthesia, to be sure) or heart-lung bypass through cross-circulation or a bubble oxygenator before the surgical procedure could even begin. Photographs of open-heart surgery published in national magazines depicted teams of ten or twelve doctors, nurses, and technicians working intently, surrounded by the technologies of the modern operating room: monitors, hoses and tubes, intense lights, instruments by the dozen. A cover story in *Time* magazine included a half-page photograph of a patient immersed in an ice bath before surgery and another photograph, in full color, showing her chest sliced open from one side to the other.[61] The surgeons acquired an image as intense and decisive wielders of these instruments. In the immediate aftermath of the surgery, some children were left dependent on a wire sticking through their skin and attached to an electronic box plugged into a wall socket. The whole technological array, including the surgeon, probably frightened many.

Arriving in early 1958, the Medtronic external pulse generator was a latecomer to this technological ensemble. While it is doubtful that Bakken intended it, his invention contrasted sharply with the equipment and procedures of open-heart surgery. Like the Apple computer of the 1980s, the 5800 had reassuring, 'friendly' features: it was white,

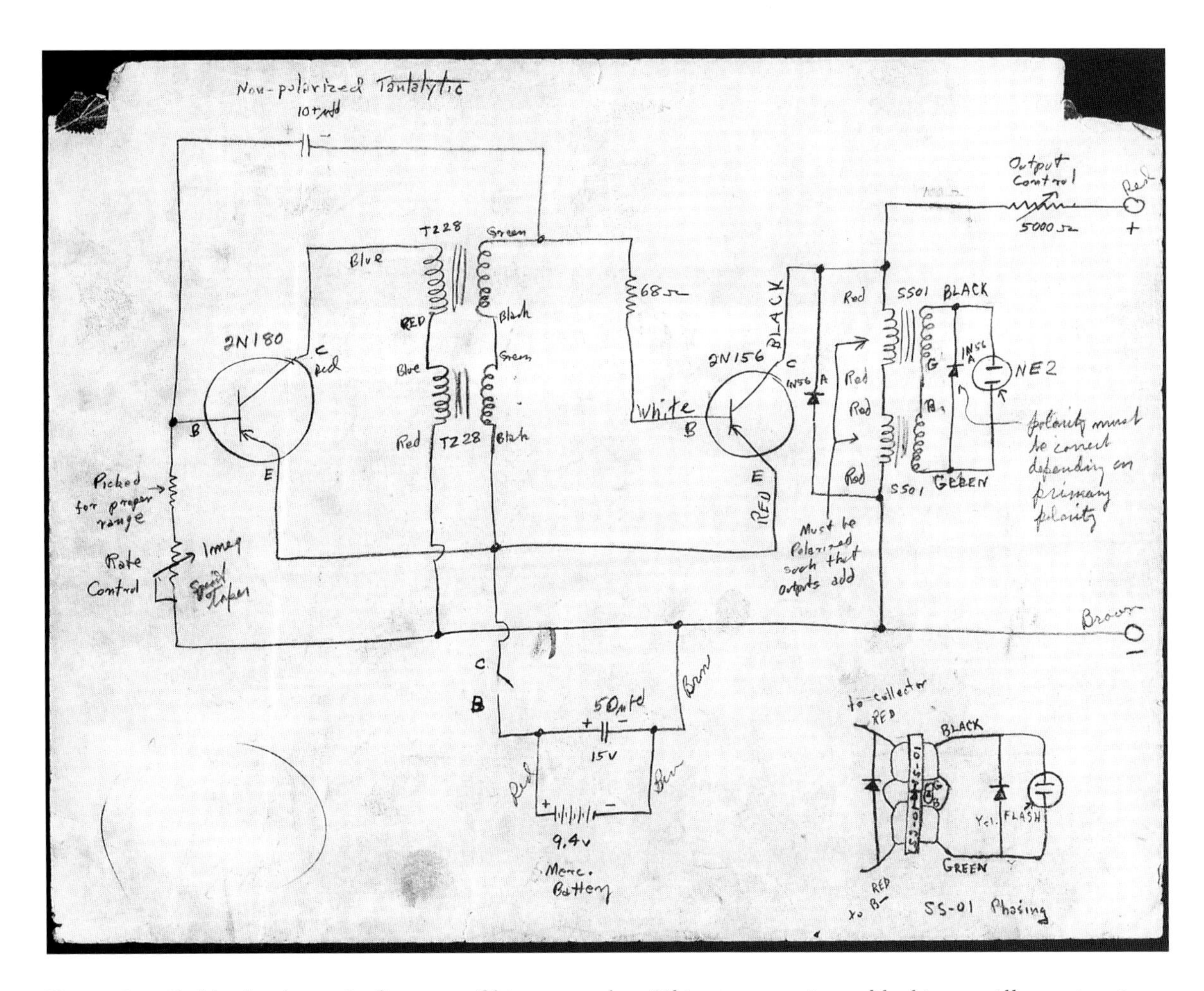

Figure 8. Bakken's schematic diagram of his pacemaker. This two-transistor, blocking-oscillator circuit was adapted from an electronic metronome described in Popular Electronics. *Courtesy of Bakken Library and Museum.*

small enough to hold in the hand, battery-powered, and as comprehensible to ordinary people as a transistor radio. And it had that blinking light.[62] The 5800 bridged the gulf between the idealized world of childhood and the somewhat frightening world of open-heart surgery. It was a reassuring technology, a token of a better future. It reflected what John Kasson has called 'the dominant popular conception of history as a steadily progressive record... .' It was a talisman, an object with a quasi-magic ability to protect the owner/wearer from harm.[63] 'I used to dread going outdoors alone,' Louise Kreher of Buffalo, N.Y., told a reporter, 'because I was always afraid I'd faint and fall. I'd go across the street to the store and then wonder why I had come [because her heart block made her dizzy and forgetful] and would feel like crying. Now I go out often; I can shop for an hour or two and only feel a little tired, nothing to worry about.'[64]

The degree to which a physician and a patient could place their confidence in Earl Bakken's little white box with the blinking red light is suggested in an anecdote told to us by Dr. Sam Hunter. Warren Mauston enjoyed his status as a celebrity patient and would happily show off his pacemaker apparatus to surgeons and cardiologists who had come to Minneapolis-St. Paul to learn about cardiac pacing and study the Medtronic 5800. He even appeared in a short promotional film produced for Medtronic in 1960. Hunter recalls that Mauston loved being in the limelight:

> And ... he allowed me ... to turn off the pacemaker and time how long before he slipped into unconsciousness... . If I set him at 60 [beats per minute] and then turned [it] off—bang—he would be O.K. for four beats. For four seconds. And then he would start to slide quickly and go unconscious or begin to twitch. And he always said he was falling back, sort of down a well or down a big barrel. And he said it wasn't unpleasant. Then I'd snap it on again, and he'd come right out of it. I did that several times. I had a lot of [ECG] tracings. I had those all over the laboratory, Mr. Mauston sliding toward eternity because I'd turned off his pacemaker.[65]

Among the early pacemaker patients, Warren Mauston stands out for his cheerful willingness to serve as an object of medical study and his determination to stay with an external pulse generator long after it had become obsolete. He contributed to the shaping of his own care and helped spread the message that it was possible to manage the heartbeat over months and years. The fact that a pacemaker recipient could rise from bed to do light housework, go bowling or dancing, work in the garden, or, in one case, leave her husband and seek more lively companionship, was big news around 1959–61, and stories appeared in national newspapers and magazines. Bakken comments that 'it was obvious that pacemakers were not only changing people physically but mentally... . They became different people with the new blood flow to their brain.'[66]

Bakken particularly recalled a patient who resided at the Veterans Administration hospital in south Minneapolis. 'They would release him every weekend to go home to Bemidji [in northern Minnesota], and he was an avid square dancer, that was his big love of life. He'd go dancing ... and he'd break his wire and then he'd retract to a slow heart rate, would have to sit down, get back to the VA, and I'd be called invariably on Mondays to come out and solder his wires back together.' Cases like this convinced him that the pacemaker was benefiting 'the whole personality of the person,' effecting a mental and spiritual as well as a physical restoration.[67] Out of such cases came the Medtronic slogan—'Toward Full Life'—and the image at the company's Web site of a human figure arising from a sickbed to a standing position. In an effort to ensure that employees remained in personal touch with patients, Bakken in later years would often arrange for a patient to visit company facilities and meet the people who had worked on his or her pacemaker. Today, Medtronic has an annual Holiday Program for employees each December at which patients carrying Medtronic implantable pacemakers, defibrillators, or other products tell about their experiences and thank the company's employees.[68]

Token of a Golden Age: the 5800, Medical Alley, and the University of Minnesota

For Earl Bakken and Walt Lillehei, and perhaps for a number of the older hands at Medtronic and in the community of pacing physicians, the 5800 has in recent years come to symbolize the heroic early days before passage of the Medical Device Amendments of 1976, the law that empowers the U.S. Food and Drug Administration [FDA] to regulate life-sustaining medical devices. A few years ago, Bakken discovered the phrase 'Ready, Fire, Aim!' To most of us, it pokes fun at the tendency to go into action before you've really planned out what you want to do. But to him, it has a rather different meaning. It means getting the product designed and built and out the door without getting bogged down in a lot of planning sessions and market projections, and especially without having to submit everything to a federal agency for approval. He is rather proud that the 5800 moved so quickly from bench testing to animal trials to managing the heartbeats of surgical patients.

Because Lillehei put the device into clinical use so quickly, many of its technical imperfections such as the need to recess the knobs and the fact (which emerged later) that the 5800 didn't function reliably in high humidity were worked out along the clinical road, not a priori. Once the 5800 had gone into everyday use, Bakken and the surgeons also discovered that the pacemaker was susceptible to electromagnetic interference from cautery machines in the operating room. As a surgeon finished up a heart operation, the cautery would momentarily make the pacemaker 'go crazy.'

And so we built in the shielding that we needed to prevent that and found out a few of those problems, and the real limits we wanted on the rate and how much else that we needed, those things were all discovered. It was such an easy way to do it: *just to do it and test in a real case, in the real world, as to where you want it.* You know you have to design it to be tough, because while the people working in surgery are wonderful people, they are not careful. Things will fall in the floor and they'll do anything they can to destroy it. Not intentionally, but that's just the way it is under the pressure of surgery. You have to build things that will stand up against these gentle little nurses that can destroy a truck. But those are things you learn by just doing it.[69]

'I think it was interesting,' Bakken has said, 'that it was only four weeks from the time that Lillehei and I talked about the need till we were using it on children. And, of course, you couldn't begin to do anything like that today, even four years! That was a time when you really could use your intelligence and your personal responsibility to do things for people ..., which we have to go overseas to do today.'[70]

When David Rhees recently examined the 'dog model' prototype with curator Ellen Kuhfeld at The Bakken Library and Museum, they found features that fit with Earl Bakken's recollection of a Ready, Fire, Aim invention process. The original box shows evidence of having been used earlier for some other purpose; the output control was probably just grabbed off the shelf since a variable resistor is built right into the knob—an arrangement most suited for infrequent adjustment; there is no calibration on that dial, and the knob is marked with a crude yellow slash from a resistor coding paint set. The device departs in several ways from the circuit diagram, apparently because Bakken sometimes didn't have the right component on hand when the time came to put it together. For example, the diagram calls for a 9.4-volt mercury battery, but the box contains a cheaper and more readily available 9-volt carbon-zinc battery.[71]

Electrostimulation has been one of the core technologies not only for Medtronic but for the medical device industry in Minnesota. In 1990 the state had about 175 firms engaged in inventing, manufacturing, and selling a wide range of advanced devices and equipment: surgical instruments and appliances, ophthalmic products, hearing aids, drug delivery systems, and heart valves. Dozens of these companies, including many that compete directly with Medtronic in various markets, were founded by former Medtronic employees. In 1984 the device manufacturers of Minnesota launched a trade association called the Medical Alley Association. The name was chosen to evoke the image of California's Silicon Valley.[72]

Many factors contributed to the rise of the medical device industry in Minnesota. Changing patterns of disease and the aging of the American population have opened business opportunities in the treatment and management of chronic diseases. The rise of third-party medical payment programs has greatly encouraged the consumption of medical services and technologies. The proliferation of medical specialties and subspecialties, many tied to specific technologies, has also encouraged the growth of the device industry.[73] Circumstances unique to Minnesota have proved important too, such as the existence of important electronics and computer firms by the 1950s, the availability of venture capital, and the internationally recognized medical education and research programs at the University of Minnesota and the Mayo Clinic, which is located 75 miles southeast of Minneapolis-St. Paul.[74]

Medical Alley represented company leaders' anxieties that the era of rapid growth in their industry based on technological innovation might be coming to an end. Many shared Bakken's concern about the Medical Device Amendments of 1976. They also cited a series of Congressional hearings that criticized some device firms for product recalls and shady marketing practices, and the efforts underway in the early 1980s to slow the growth of health-care spending, particularly in the Medicare program.[75] Spokesmen for Medical Alley (including Bakken, who served as president of the organization for a time in the 1980s) tried to make the case that delays in the FDA approval process seriously impeded the effort of Minnesota-based firms, including Medtronic, to remain competitive in the global market for medical devices; they raised the spectre of a 'brain drain' of talented medical researchers and bioengineers from the U.S. to western Europe. The Safe Medical Devices Act of 1990, which expanded the regulatory scope of the FDA and empowered the agency to seek civil penalties against device manufacturers, further alarmed leaders of the Minnesota device companies.[76]

Shortly after the formation of Medical Alley, the University of Minnesota announced a new Bakken Chair in Biomedical Engineering funded with a gift of $2 million from Medtronic.[77] Then in the 1990s, the University launched a development campaign to endow a Biomedical Engineering Institute that would promote collaborative research between university-based scientists and engineers and the biomedical device companies of Minnesota. To the surprise of no one, Earl Bakken and Walt Lillehei served as honorary co-chairmen for this campaign. The interests of the companies of Medical Alley and of planners and fundraisers at the University of Minnesota thus had very clearly converged. For the kickoff of the public phase of the campaign in 1997, a series of media events commemorated the fortieth anniversary of the invention of the transistor pacemaker. Both the surgeon and the inventor were on hand to participate in interviews, a press conference, and appearances on public TV and radio. Soon afterward, a plaque commemorating the invention of the 5800 was mounted on the wall of the former operating room where Bakken and Lillehei had worked together.[78]

The dual effort to raise the profile of the medical device and health care industries in Minnesota and to generate financial and political support for the new Biomedical Engineering Center at the University converged on a common creation story, the account of the Bakken-Lillehei collaboration in 1958, and a common symbol, the Medtronic 5800, 'the world's first wearable, battery-powered pacemaker.'[79] The 5800 pacemaker represented a successful 'academic-industrial technology transfer link'; it was the obligation of Minnesotans of the 1990s to ensure (by supporting the Biomedical Engineering Center) 'that when a future C. Walton Lillehei has an idea for a medical device, there will be an Earl Bakken nearby to make sure the job is done right.'[80]

The 5800 at Medtronic Today

When they examined the original 'dog model,' Rhees and Kuhfeld noticed that someone had glued a strip of velcro along the bottom edge of the artifact. This indicates that it had been mounted for display at some time—we think probably for internal Medtronic exhibitions or for medical trade shows.[81] More recently Medtronic has had several replicas of the dog model made for display at various corporate facilities around the world. In June 1998, Earl Bakken presented one of these replicas to managers and employees at the new Medtronic Europe headquarters facility in Tolochenaz, Switzerland.

Earl Bakken made it a practice throughout his years at Medtronic to meet with every newly hired employee. Even today, though officially retired, he occasionally visits company facilities to talk to new people. This orientation session has evolved into a somewhat ceremonial affair now called the Mission/Medallion Ceremony. Bakken discusses at length the six-point Mission Statement that he wrote and debated with his board of directors in 1961–62. He also reviews the history of the company, including his invention of the 5800. Sometime in the early 1990s, he began to use the prototype at these orientation meetings. More recently, he has switched to a replica of the 5800 and keeps the original at The Bakken Library and Museum for safekeeping. If he is talking to new employees at Medtronic headquarters near Minneapolis, Bakken wheels in a cartful of vacuum-tube apparatus to show what he avoided by building the small transistorized pacemaker. He uses the prototype as a token of a simpler, freer era that contrasts with today's highly regulated world. He mentions that it took only four weeks to get the 5800 into clinical use, and that gets a big ironic laugh from the employees. Bakken understands, of course, that we can't go back to that era, but he uses the prototype to encourage the Ready, Fire, Aim approach, and perhaps also to undermine the natural conservatism of a now very large company. At the end of the ceremony, each employee is given a medallion containing the company motto, 'Toward Full Life', and the image of a man rising from a gurney. The 5800, the Mission Statement, the Holiday Program, and Earl Bakken himself represent continuity with the company's past; they assert that at bottom, Medtronic is the same kind of company that it was in the early days.[82]

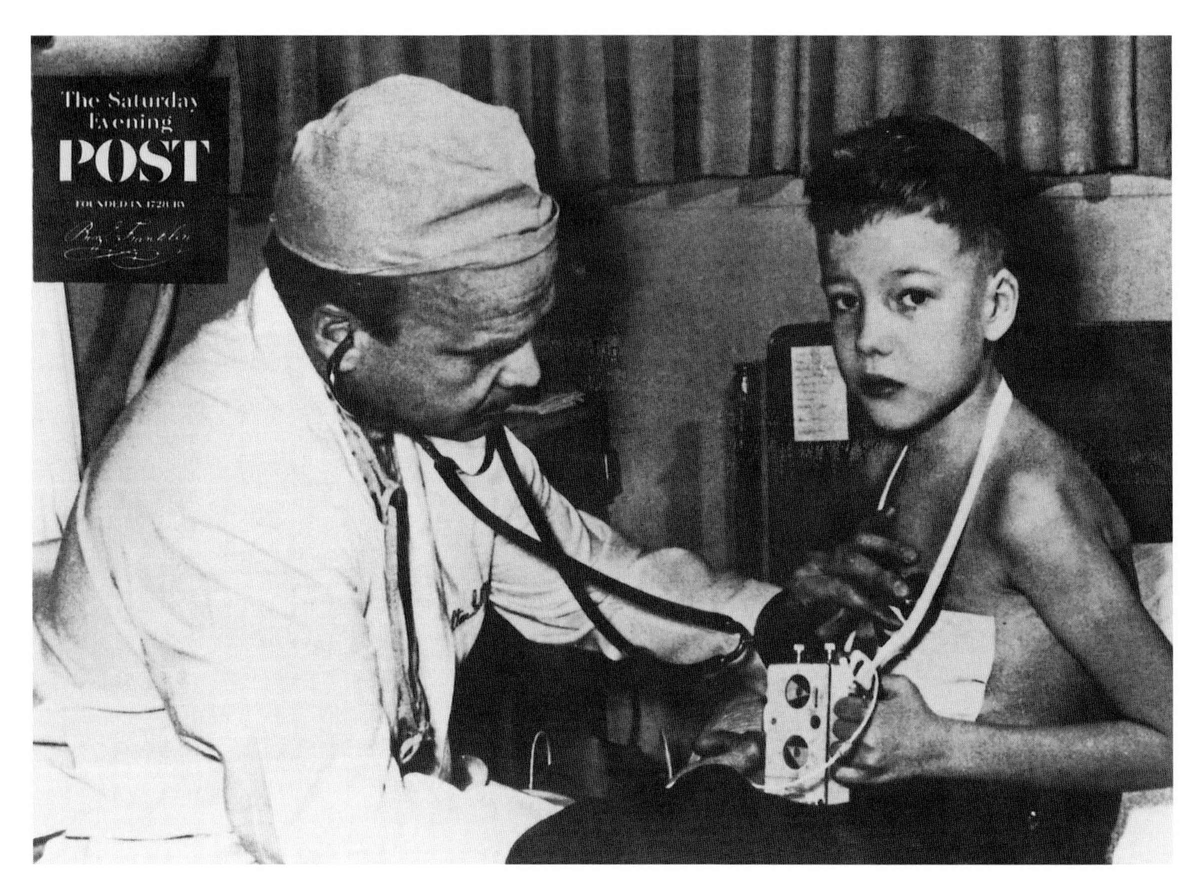

Figure 9. Dr. C. Walton Lillehei, a pioneer in open-heart surgery, examines patient David Williams, who is wearing a Medtronic 5800 pacemaker. This appeared in the March 4, 1961 issue of Saturday Evening Post. *Courtesy of Medtronic, Inc.*

Over the years, the stories of the 1950s—the garage behind the Hermundslie house, the blue babies at the University of Minnesota, Dr. Lillehei and the myocardial pacing wire, Bakken's inventing the external pulse generator using a metronome circuit—have taken on a mythic character; and as a longtime Medtronic board member remarked recently, 'companies live on myths.' A business writer who studied Medtronic in the 1980s found that 'everyone knows the story. The Garage ... symbolized an unfettered state where technical genius and creativity could be applied for the betterment of mankind.' Stories of the early days also remind employees that the company's success depended on close collaboration between engineers and daring, innovative physicians. The 5800 pulse generator has helped Bakken make the link, for himself and for Medtronic employees, between those heroic early days and the much larger and necessarily more formally structured company that emerged later.[83]

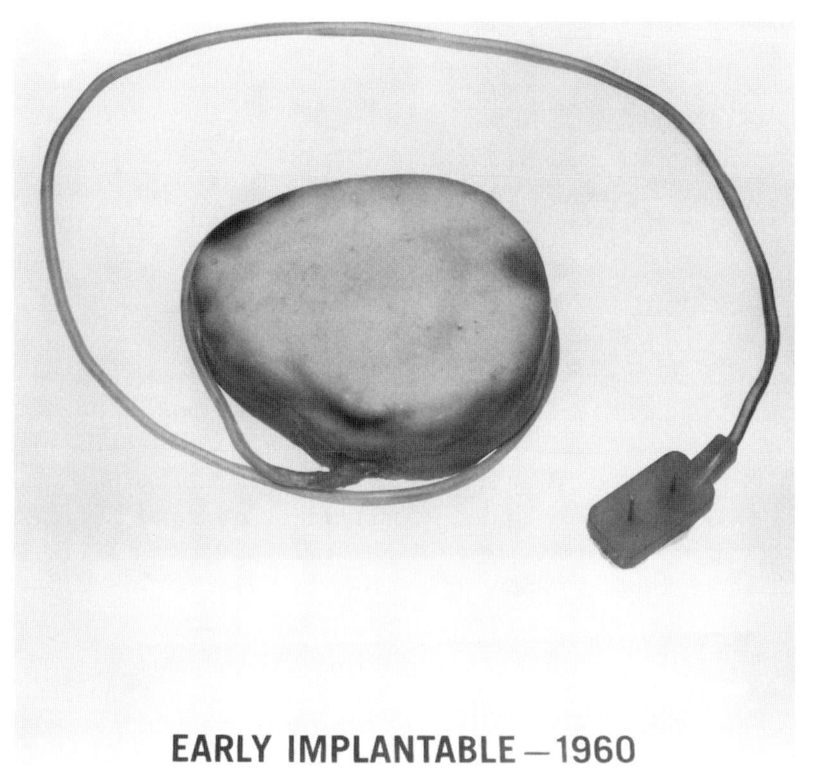

Figure 10. The Chardack-Greatbatch implantable pacemaker, 1960, which Medtronic obtained the rights to produce. The pulse generator is encapsulated in silicone rubber and is fitted with a Hunter-Roth platform electrode.
Courtesy of Medtronic, Inc.

When physicians come to Medtronic headquarters, they usually visit the Bakken Education Center to attend short courses on heart rhythm disorders or to learn about the newest Medtronic pacemakers. In the lobby just outside the doors to the auditorium is a glass display cabinet that extends from floor to ceiling and holds every major pacemaker model that the company has produced.[84] The 5800 is there too (in replica), looking ancient and ungainly compared to the tiny, sleek implantable pacers of the 1990s with their titanium shells. Cardiac pacing for slow heartbeats has probably reached technological maturity, but the cycle of innovation that the 5800 inaugurated continues because physicians and manufacturing firms have applied the basic idea of electrostimulation to other disorders. Pacemakers have become tiny implanted computers able to diagnose heart rhythm disorders and select an appropriate kind of pacing on their own. Implanted defibrillators can deliver a series of increasingly powerful shocks to terminate ventricular or atrial fibrillation, then pace the heart until it resumes beating on its own. Implanted stimulators now treat a range of neurological disorders including chronic pain, urinary incontinence, epilepsy, essential tremor, and sleep apnea. Earl Bakken recently remarked to David Rhees, 'If the electricity in the body stops flowing, then life is over. That's why with

Medtronic now, our devotion is to electrical stimulation. We're just beginning to scratch the surface of what can be done with electrical stimulation, many times combined with some chemical, but mainly the basis of our work is the electrical. That's the way it'll be in the future.'[85]

Conclusion

Like any artifact of significance, the 5800 has meant different things to different groups, and these meanings have evolved over time as circumstances changed. We have seen that initially the 5800 was perceived by its inventors as a useful but incremental technological advance intended to treat a side effect of heart surgery that affected only a small percentage of a small group of patients. Bakken has said that he had no inkling at the time that 'history was being made' when he built the prototype. That perception changed beginning in the early 1960s with increasing use of the 5800 and with the development of the Hunter-Roth bipolar electrode and the earliest completely implantable pacemakers. As physicians improved their understanding of heart block and other disorders of heart rhythm, and realized that these disorders were by no means uncommon in elderly people, the market for transistorized pacemakers expanded enormously. As Bakken and Medtronic successfully pursued this market, the modest 5800 gradually acquired greater significance. Instead of just another 'special,' the device came to be remembered as the golden opportunity that enabled a struggling company to become the leader of a new, high-tech industry and gave thousands of Medtronic employees over the years the satisfaction of having helped sick people acquire a new lease on life.

Lillehei's interests, of course, focused more broadly on heart surgery—he went on to contribute important innovations in heart valves and many other areas—but he too came to view the 5800 as an important technological watershed. It represented to him, as to Earl Bakken, the remarkable fruits of interdisciplinary collaboration, exemplifying the wedding of medicine and technology that produced many successful therapeutic and diagnostic tools after World War II. But the 5800 meant more to Lillehei and Bakken than just a symbol of postwar collaborative R&D. As was clear to David Rhees during his video interview with the two of them, the little white box also became the emblem of long-lasting friendship and mutual respect.[86]

The meaning of the 5800 to other physicians is more difficult to document, but our limited evidence suggests that doctors' initial response was somewhat wary. In the late 1950s, when electronic devices were just beginning to intrude into the practice of medicine, cardiologists were perhaps a bit intimidated by the pacemaker. To general practitioners, especially, prescribing the externally-worn 5800 probably seemed somewhat risky, both medically and legally. Perhaps doctors also felt uneasy with being so dependent on engineers and technicians. On the other hand, the elegance of the 5800 with its sleek, white plastic housing and its miniaturized, transistorized technology must have seemed very attractive to physicians, in spite of any concerns they may have had. To wield a device that with the flick of a switch could take complete control of a patient's heart and instantly restore it to a normal rhythm seemed the very essence of up-to-date scientific medicine. In this sense, the 5800 and its many successors served to affirm and publicly display the legitimacy of the medical profession's progressivist outlook.

The meaning of the 5800 to patients and the broader public is also difficult to determine, but our admittedly anecdotal evidence suggests that patients responded enthusiastically. Adult wearers of the device and the media hailed it as yet another example of the achievements of modern medical science. The 5800 was uniquely reassuring by virtue of its small size, apparent simplicity, and red blinking light. It was a relatively non-threatening technology that restored patients to a sustainable lifestyle, and in some instances it may have had profound emotional and spiritual effects on its wearers.

In 1999 Earl Bakken and Medtronic celebrated the fiftieth anniversary of the founding of the company in a northeast Minneapolis garage. The 5800 played an important role in the celebration, which included the publication of Bakken's autobiography. As Medtronic expands rapidly and tries to integrate dozens of newly acquired small companies and their employees, it will look to the 5800 as a symbol of continuity and particularly of the company's desire to remain close to physicians and to patients. Medtronic's officers turn to history for a reason: historical symbols and ceremonies play an important part in the ongoing effort to acculturate new employees and to create a unified set of values that will hold the company together.[87]

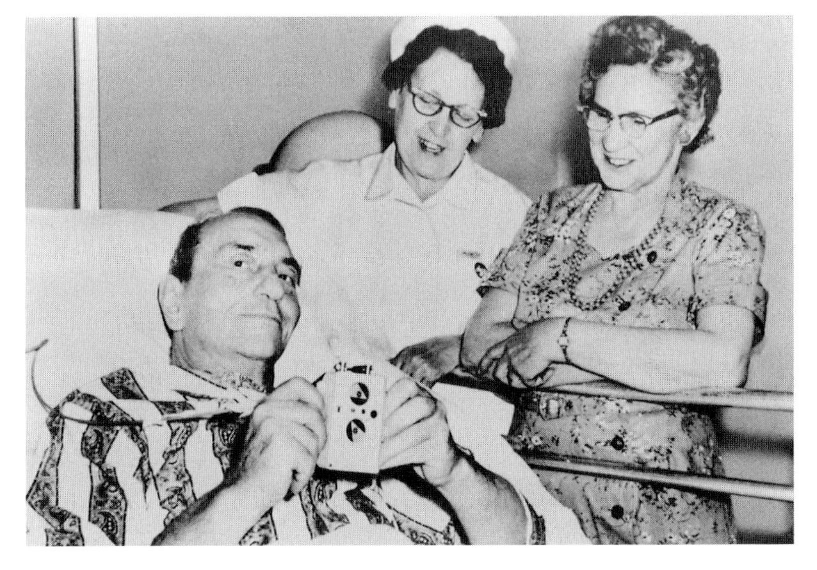

Figure 11. Warren Mauston of St. Paul, Minnesota, lived from April 1959 to October 1966 with a Medtronic external pulse generator and a Hunter-Roth bipolar electrode. Courtesy of Medtronic, Inc.

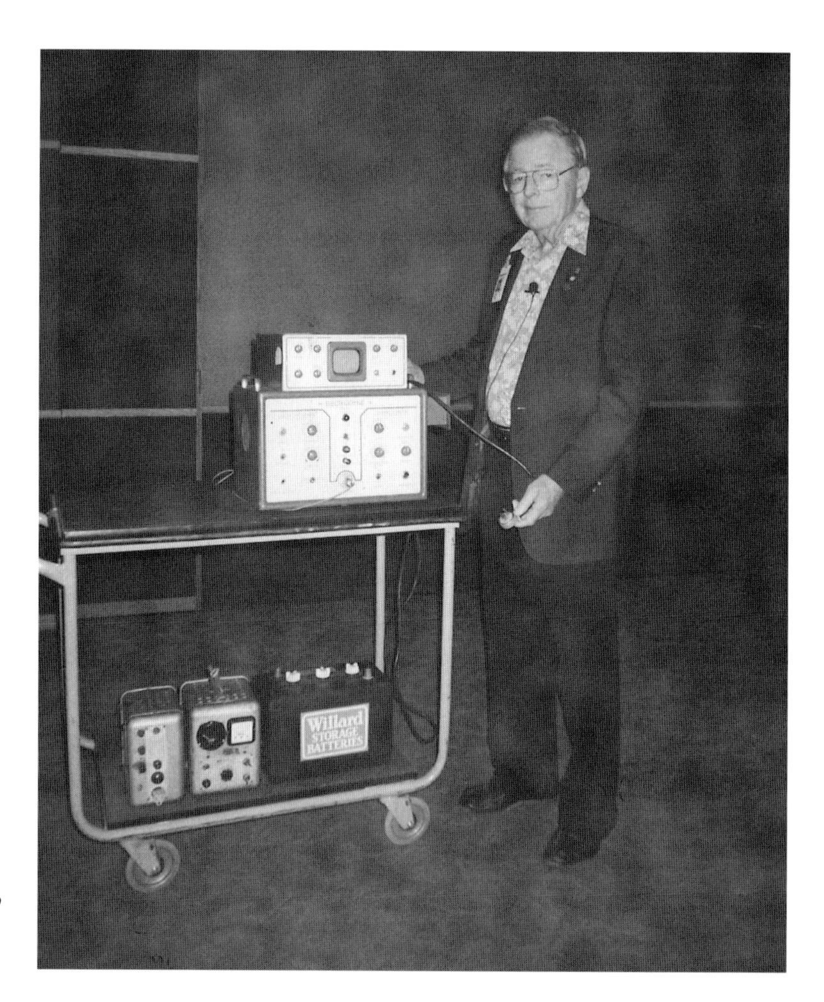

Figure 12. Earl Bakken in 1998, during a Mission/Medallion Ceremony for new Medtronic employees. He uses the cart of apparatus (an AC-powered pacemaker/ECG unit, car battery, recharger, and inverter) to show what he avoided by switching from vacuum tubes to transistors. Courtesy of Bakken Library and Museum.

Elsewhere in Minneapolis, the original 5800 prototype is now on display at The Bakken Library and Museum. The Bakken, as it is called, is a nonprofit institution that Earl Bakken founded in 1975, whose mission is to promote understanding of the history, cultural context, and applications of electricity and magnetism in the life sciences. There the 5800 has found a place of honor in a new wing of the museum that was completed in 1999, where the prototype is displayed in a lobby adjacent to a bronze bust of Bakken, made from the same mold as the full-sized statue at Medtronic's headquarters. In this setting, the 5800 serves a broader audience and purpose, exemplifying the kind of creative, innovative thinking that The Bakken hopes to inspire in the students and other visitors it serves. Some of these young people may one day invent their own versions of Earl Bakken's little white box.

Thanks to Earl E. Bakken, C. Walton Lillehei, Ronald T. Hagenson, Ellen Kuhfeld, and staff members of the Minnesota Historical Society for their assistance with this essay.

Notes

1. Medtronic is the world leader in market share in pacemakers and is a close second in implantable defibrillators. The company also manufactures heart valves, angioplasty catheters, and coronary stents; implantable neurostimulation devices for pain management, sleep apnea, and the treatment of essential tremor in Parkinson's disease; neurosurgery products; implantable drug delivery systems; and other products. For the fiscal year ending on 30 April 1998, Medtronic had 14,000 employees worldwide and net sales of $2.605 billion, of which 64% was generated by its Cardiac Rhythm Management business (pacemakers, defibrillators, and related equipment): Medtronic annual report, 1998.
2. This was not, of course, a working replica of the pacemaker but a sculpted shape that evoked the general appearance of the artifact.
3. Earl E. Bakken, interview by David J. Rhees, 10 January 1997, transcript in interviewer's possession. The surgeon, C. Walton Lillehei, called this early prototype the 'dog model,' and the name has stuck.
4. Christopher P. Toumey, *Conjuring Science: Scientific Symbols and Cultural Meanings in American Life* (New Brunswick, N.J.: Rutgers Univ. Press, 1996): 48–49 et passim. Toumey's leading example, having to do with meanings imputed to nuclear power, involves a technological system.
5. Trevor J. Pinch and Wiebe E. Bijker, 'The Social Construction of Facts and Artifacts: or how the Sociology of Science and the Sociology of Technology might Benefit each other,' in *The Social Construction of Technological Systems*, ed. Bijker, Thomas P. Hughes, and Pinch (Cambridge, MA, 1987): 17–50; Bettyann Holtzmann Kevles, *Naked to the Bone: Medical Imaging in the Twentieth Century* (New Brunswick, 1996).
6. Melvin Kranzberg, 'Technology and History: "Kranzberg's Laws",' *Technology & Culture* 27 (July 1986): 544–560.
7. Louis Galambos with Jane Eliot Sewell, *Networks of Innovation: Vaccine Development at Merck, Sharp & Dohme, and Mulford, 1895–1995* (Cambridge, Eng., 1996).
8. Charles E. Rosenberg and Janet Golden, eds., *Framing Disease: Studies in Cultural History*, ed. Rosenberg and Janet Golden (New Brunswick, 1992), especially the essays by Steven J. Peitzman and Christopher Lawrence. For physicians' framing of new heart rhythm disorders treatable with the pacemaker, see Kirk Jeffrey, 'Pacing the Heart: Growth and Redefinition of a Medical Technology, 1952–1975,' *Technology & Culture* 36 (July 1995): 583–624.
9. Pacemaker Market shows Slight Growth,' *Cardiovascular Network News* 4 (November 1997): 7; confidential industry sources. The figure $2.5 billion in sales covers pacemaker equipment only and does not include implantable defibrillators. A pacemaker speeds the heartrate when the rate is too slow to maintain an adequate circulation of blood to the body's organs, while a defibrillator terminates episodes of ventricular fibrillation, random electrical activity in the ventricles that prevents any organized heartbeat. A pacemaker works by firing tiny electrical impulses into the right atrium and/or right ventricle of the heart at an appropriate rate; a defibrillator gives the heart a high-energy shock that terminates all electrical activity and permits an organized heartbeat to resume. In 1996, physicians implanted about 142,000 pacemakers in the U.S., 170,000 in Europe, and 75,000 elsewhere, counting first-time and replacement implants. The three Twin Cities firms held about 80% of the world market for pacemakers and virtually the entire world market for implantable defibrillators.
10. 'Electronics Boom Spreads to medicine,' *Medical World News* 2 (18 August 1961): 15–17; Charles K. Friedberg, *Medical Electronics in Cardiovascular Disease* (New York: Grune & Stratton, 1963); Lawrence Lessing, 'The Transistorized M.D.,' *Fortune* 68 (September 1963): 131+; 'Electronics helps to heal the sick,' *Business Week*, 19 September 1964: 75–78; Hughes Day, 'An intensive coronary care area,' *Diseases of the Chest* 44 (October 1963): 423–427; Paul M. Zoll, 'The cardiac monitoring system' [interview], *Medical World News* 186 (2 November 1963): 33–36; W. Bruce Fye, *American Cardiology: The History of a Specialty and Its College* (Baltimore, 1996), pp. 176–181, 250–254.
11. Paul M. Zoll, 'Resuscitation of the Heart in Ventricular Standstill by External Electric Stimulation,' *New England Journal of Medicine* 247 (13 November 1952): 768–771; Kirk Jeffrey, 'The Invention and Reinvention of Cardiac Pacing,' *Cardiology Clinics* 10 (November 1992): 561–571.
12. Leonard G. Wilson, *Medical Revolution in Minnesota: A History of the University of Minnesota Medical School* (St. Paul, 1989): chapter 20.
13. Tetralogy of Fallot is a congenital malformation of the heart named for the French physician who had first described the condition in 1888: William F. Friedman, 'Congenital Heart Disease in

Infancy and Childhood,' in *Heart Disease: A Textbook of Cardiovascular Medicine*, ed. Eugene Braunwald (4th edn.; Philadelphia, 1992): 935.

14. The three technologies were controlled hypothermia or lowering the body temperature so as to slow the metabolism and reduce the body's need for oxygenated blood, thereby giving the surgeon a few minutes to work in the slowly beating heart; cross-circulation, in which an artery and a vein from the surgical patient were connected to the circulatory system of a parent, whose heart then pumped blood for both bodies during surgery; and the bubble oxygenator, the first practical and reliable heart-lung machine. A group in Toronto had conceived of the hypothermia technique in 1950 and it was under study in several surgical programs besides the one at Minnesota: Wilfred A. Bigelow, John C. Callaghan, and John A. Hopps, 'General Hypothermia for Experimental Intracardiac Surgery,' *Annals of Surgery* 132 (September 1950): 531–539. Surgeons at Minnesota were, however, the first to use hypothermia clinically in open-heart operations (September 1952). Cross-circulation had been tried with laboratory animals in England but was first used clinically at Minnesota. See Vincent L. Gott, 'C. Walton Lillehei and Total Correction of Tetralogy of Fallot,' *Annals of Thoracic Surgery* 49 (February 1990): 328–332; Wilson, *Medical Revolution in Minnesota*, pp. 488–517; C. Walton Lillehei and Leonard Engel, 'Open-Heart Surgery,' *Scientific American* 202 (February 1960): 76–90; and Stephen L. Johnson, *The History of Cardiac Surgery, 1896–1955* (Baltimore, Md.: Johns Hopkins Univ. Press, 1970), 113–119. Many surgeons believed that cross circulation was too risky for everyday use; one visitor told C. Walton Lillehei that he was the first surgeon who had ever devised an operation with a potential for a 200% mortality rate: C. Walton Lillehei, interview by Kirk Jeffrey, 25 July 1990, NASPE [North American Society of Pacing and Electrophysiology] Oral History Archive, Natick, MA.
15. C. Walton Lillehei, Herbert L. Warden, Richard E. DeWall, et al., 'Cardiopulmonary By-pass in Surgical Treatment of Congenital or Acquired Heart Cardiac Disease,' *Archives of Surgery* 75 (December 1957): 928–945; Wilson, *Medical Revolution in Minnesota*, pp. 508–516. As Wilson shows, an improved oxygenator known as a sheet oxygenator (also pioneered at Minnesota) supplanted the bubble oxygenator within a few years.
16. For popular worship of heart surgeons see, e.g., 'Stopped Heart Operation,' *Reader's Digest* 69 (August 1956): 29–33; 'Rare, Revealing Look into a Beating Heart,' *Life* 41 (3 December 1956): 127–128; 'Inside the Heart: Newest Advances in Surgery,' *Time* 69 (25 March 1957): 66–77.
17. Earl E. Bakken, interview by William Swanson, 22 December 1992, transcript at The Bakken Library and Museum, Minneapolis, Minn. Lillehei said, '50 to 75 volts was intolerable… . It takes that much to stimulate the heart through the chest. Getting a shock like that fifty, sixty times a minute is torture. With some of the infants, we were able to restrain them so they wouldn't tear [the chest electrodes] off, but they would develop blisters and ulcers [beneath the electrodes] in four to five days. So that was totally inadequate': Lillehei interview by Jeffrey; C. Walton Lillehei, Morris J. Levy, Raymond C. Bonnabeau, et al., 'Direct Wire Electrical Stimulation for Acute Postsurgical and Postinfarction Complete Heart Block,' *Annals of the New York Academy of Sciences* 111 (11 June 1964): 938–949.
18. Herbert E. Warden, 'C. Walton Lillehei: Pioneer Cardiac Surgeon,' *Journal of Thoracic and Cardiovascular Surgery* 98 (November 1989): 33–45; C. Walton Lillehei, videotaped conversation with Earl E. Bakken, 10 June 1977, 'Pioneers in Pacing' videotape series, The Bakken Library and Museum.
19. It will be recalled that the heart in dogs and human beings has two small upper chambers, the atria, and two main pumping chambers, the ventricles. The right ventricle pumps blood to the lungs while the left ventricle pumps oxygenated blood to the body. The myocardial wire was attached to the outer surface of one of the ventricles. Warden, 'C. Walton Lillehei'; Vincent L. Gott, interview by Kirk Jeffrey, 2 May 1997, transcript in interviewer's possession. Lillehei recalled, 'They created some heart block, put the wire in the heart, and lo and behold, one or two volts, five to ten milliamps … drove the heart beautifully! Any rate that you'd set. And obviously, one to two volts was totally imperceptible to the animal. And there was no FDA [Food and Drug Administration] in those days. We ran some animals—I don't know, ten, fifteen—and it just worked beautifully': Lillehei interview by Jeffrey.
20. Ibid.; Wilson, *Medical Revolution in Minnesota*, pp. 517–519.
21. [Northern States Power Co.,] 'Report on October 31, 1957 Major Disturbance,' typescript, copy courtesy of Ronald T. Hagenson.
22. C. Walton Lillehei in Lillehei and Earl E. Bakken, videotaped interview by David J. Rhees, 9 September 1997, Pioneers in Minnesota Medical Device Industry Oral History Series (Minnesota Historical Society, St. Paul, Minn.).

23. Medtronic went public in the 1960s. Palmer Hermundslie, a diabetic, remained actively involved in managing the company but died in 1970 at age 51.
24. Bakken interview by Swanson.
25. Earl E. Bakken, interview by Kirk Jeffrey, 23 May 1990, transcript in interviewer's possession; Bakken interview by Swanson; Art Detman, 'The Reluctant Millionaire,' *Dun's Review* 103 (March 1974): 12–19 at 13–14; Lillehei conversation with Bakken, 1977, 'Pioneers in Pacing' Series; Louis E. Garner, 'Five New Jobs for Two Transistors,' *Popular Electronics* 4 (April 1956): 54–59.
26. The Medtronic 5800 was first formally described in print in 1960: C. Walton Lillehei, Vincent L. Gott, Paul C. Hodges, Jr., et al., 'Transistor Pacemaker for Treatment of Complete Atrioventricular Dissociation,' *Journal of the American Medical Association* 172 (30 April 1960): 2007. See also the comments in an early company history, 'Medtronic: a Look Back' (Minneapolis, Minn.: Medtronic, Inc., 1970): 6–7; copy in Medtronic Information Resources Center, Fridley, Minn.
27. Bakken interview by Rhees, 9 September 1997. At the Karolinska Institute in Stockholm, surgeon Åke Senning and engineer Rune Elmqvist built an external pacemaker of similar design around 1958. Senning had visited Minnesota to study Lillehei's surgical techniques and had observed the use of the myocardial pacing wire, but this was in 1957 before Bakken had invented the Medtronic 5800. The Swedish external pacemaker was commercialized in 1960 by Elmqvist's employer, the Swedish firm Elema-Schönander: Elmqvist et al., 'Artificial Pacemaker for Treatment of Adams-Stokes Syndrome and Slow Heart Rate,' *American Heart Journal* 65 (June 1963): 731–733; Senning, 'Cardiac pacing in retrospect,' *American Journal of Surgery* 145 (June 1983): 734.
28. Lillehei interview by Jeffrey, 1990; Bakken interview by Rhees, 10 January 1997.
29. Bakken interview by Rhees, 10 January 1997.
30. Wilson, *Medical Revolution in Minnesota*, 523; Vincent Gott, 'C. Walton Lillehei and his Trainees: One Man's Legacy to Cardiothoracic Surgery,' *Journal of Thoracic and Cardiovascular Surgery* 98 (November 1989): 846–851; Lillehei et al., 'Transistor Pacemaker.' See also 'Heart Timer: Minnesota Reports on Use of Electronic Pacemaker,' New York *Times*, 1 May 1960.
31. Earl E. Bakken, interview by David J. Rhees, 1 February 1998, Pioneers in Minnesota Medical Device Industry Oral History Series, Alan D. Bernstein and Victor Parsonnet, 'Survey of Cardiac Pacing and Defibrillation in the United States in 1993,' *American Journal of Cardiology* 78 (15 July 1996): 195.
32. Howard Bird, 'An Affair of the Heart,' *New England Journal of Medicine* 326 (13 February 1992): 487–488; Mary Moore Free, 'The Heart of the Matters of the Heart,' *American Journal of Cardiology* 78 (15 July 1996): 217–218; Lynn Payer, *Medicine and Culture: Varieties of Treatment in the United States, England, West Germany, and France* (New York, 1988), pp. 74–75, 79–85. Two recent mass-market books have reasserted some of the older notions about the heart: Paul Pearsall, *The Heart's Code: Tapping the Wisdom and Power of Our Heart Energy* (New York, 1998) and Claire Sylvia, *A Change of Heart: A Memoir* (New York, 1998).
33. See, e.g., Gerald Geison, *Michael Foster and the Cambridge School of Physiology* (Princeton, 1978); Robert G. Frank, Jr., 'The Telltale Heart: Physiological Instruments, Graphic Methods, and Clinical Hopes 1854–1914,' in *The Investigative Enterprise: Experimental Physiology in Nineteenth-Century Medicine*, ed. William Coleman and Frederic L. Holmes (Berkeley and Los Angeles, 1988), pp. 211–290; W. Bruce Fye, 'A History of Cardiac Arrhythmias,' in *Arrhythmias*, ed. John A. Kastor (Philadelphia, 1994): 1–24.
34. For comments on the interventionist attitude toward the human heart in American medical culture, see, *inter alia*, Payer, *Medicine and Culture*, pp. 23–34, 124–131, 149–152; Bruce F. Waller, '"Crackers, Breakers, Stretchers, Drillers, Scrapers, Shavers, Burners, Welders and Melters"—The Future Treatment of Atherosclerotic Coronary Artery Disease? A Clinical-Morphologic Assessment,' in *An Era in Cardiovascular Medicine*, ed. Suzanne B. Knoebel and Simon Dack (New York: Elsevier, 1991), pp. 177–194; and Fye, *American Cardiology*.
35. The Medtronic company motto is 'Toward Full Life.' For an example of a history of cardiac pacing that implicitly uses the concept of progress, see Robert D. Gold, 'Cardiac Pacing—From Then to Now,' *Medical Instrumentation* 18 (January-February 1984): 15–21.
36. For a recent discussion of the doctrine of technological determinism, see *Does Technology Drive History? The Dilemma of Technological Determinism*, ed. Merritt Roe Smith and Leo Marx (Cambridge, MA, 1994), particularly the introductory essay by Smith, 'Technological Determinism in American Culture.' For applications to medical technology, see Fye, *American Cardiology*, and

Stuart S. Blume, *Insight and Industry: On the Dynamics of Technological Change in Medicine* (Cambridge, MA, 1992).

37. The importance of non-technical factors is evident in the way pacemakers are marketed to physicians and hospitals; for example, manufacturers give their pacemakers distinctive model names—CyberLith, Dash, Activitrax, Vigor—and publish multi-page advertisements that feature pictures of the devices. The advertising has the effect of turning them into objects of desire while downplaying their functional characteristics.
38. Lillehei interview by Rhees, 9 September 1997. We speculate further that for some heart surgeons, the transistorized pacemaker with an implantable wire pointed the way toward artificial organs. The Society for Artificial Internal Organs began to publish its annual *Transactions* in 1955; Willem Kolff and Tetsuze Akatsu inaugurated their experimental work toward an artificial heart at the Cleveland Clinic in 1957. See Thomas A. Preston, 'The Artificial Heart,' in Diana B. Dutton, *Worse than the Disease: Pitfalls of Medical Progress* (Cambridge, Eng., 1988), pp. 91–126.
39. Kirk Jeffrey, 'The Next Step in Cardiac Pacing: The View from 1958,' *PACE* 15 (June 1992): 961–967.
40. In May 1976, Congress passed and President Gerald Ford signed into law the Medical Device Amendments to the Food, Drug, and Cosmetic Act of 1938, the legislation that created the Food and Drug Administration. Under the terms of the new law, cardiac pacemakers, as life-sustaining technologies, became subject to 'premarket approval' from the FDA. For a brief introduction see Michael S. Baram, 'Medical Device Legislation and the Development and Diffusion of Health Technology,' in *Technology and the Quality of Health Care*, ed. Richard H. Egdahl and Paul M. Gertman (Germantown, Md.: Aspen Systems Corp., 1978), 191–197. As Earl Bakken and C. Walton Lillehei are well aware, their technology for postsurgical pacing would have been subjected to months, perhaps years of animal and clinical trials before commercial release had they invented it after 1976 instead of in 1957–58.
41. Gold, 'Cardiac Pacing'; Wilson Greatbatch to Kirk Jeffrey, 2 March 1990; Bakken interview by Jeffrey.
42. Annetine Gelijns and Nathan Rosenberg, 'The Dynamics of Technological Change in Medicine,' *Health Affairs* 13 (Summer 1994): 31.
43. Kenneth E. Warner, 'A "Desperation-Reaction" Model of Medical Diffusion,' *Health Services Research* 10 (Winter 1975): 369–383; Payer, *Medicine and Culture*, pp. 124–131.
44. Bakken recalls that the Medtronic booth at a cardiology convention prominently featured the photograph from the *Saturday Evening Post* showing Lillehei and a surgical patient who was wearing a 5800 pacemaker (see Figure 9): Bakken interview by Rhees, 1 February 1998.
45. Bakken interview by Jeffrey. In September 1960, Medtronic signed a license agreement to manufacture and market an implantable pacemaker invented in Buffalo, N.Y., by engineer Wilson Greatbatch and thoracic surgeon William M. Chardack. The firm's close association with Chardack and Greatbatch until 1968 supplemented and eventually supplanted its earlier reliance on Lillehei and the University of Minnesota. The Medtronic annual report for 1982 had a lengthy discussion of company relationships with clinical researchers. Of the four M.D.s on the Medtronic board today, one was a surgeon and the other three are academic administrators; none was a pacemaker specialist.
46. Friedman, 'Congenital Heart Disease in Infancy and Childhood,' 888. Tetralogy of Fallot represents about ten percent of all forms of congenital heart disease: ibid., 935. In the 1950s, it was commonly said that about 50,000 infants were born annually with congenital heart defects; e.g., 'Inside the Heart,' p. 71.
47. Jeffrey, 'The Next Step in Cardiac Pacing.'
48. William M. Chardack, 'Recollections, 1958–1961,' *PACE* 4 (September-October 1981): 592. For a contemporary discussion of complete heart block, see John C. Rowe and Paul Dudley White, 'Complete Heart Block: A Follow-Up Study,' *Annals of Internal Medicine* 49 (August 1958): 260–269.
49. As Gelijns and Rosenberg have observed, 'there is a greater elasticity of demand for medical services than is commonly believed; more precisely, a downward shift in supply may bring about an outward shift in demand, with an ultimate increase in total expenditures': Gelijns and Rosenberg, 'Dynamics of technological change in medicine,' p. 39. This recalls Say's Law, 'supply creates its own demand.'
50. Norman Roth, interview by Kirk Jeffrey, 29 August 1990, NASPE Oral History Archive. Samuel Hunter, Roth's collaborator on the pacing lead, believes that Roth formulated the idea of long-term cardiac pacing with the Medtronic 5800 earlier than anyone else at Medtronic: Samuel W. Hunter, interview by Kirk Jeffrey, 10 January 1997, NASPE Oral History Archive.

51. Samuel W. Hunter, Norman A. Roth, et al., 'A Bipolar Myocardial Electrode for Complete Heart Block,' *Journal-Lancet* 79 (November 1959): 506–508. Experience later showed that the platform electrode had an unacceptably high failure rate from broken wires. After its use in some of the earliest long-term cardiac pacing, it was superseded in 1961 by new lead configurations.
52. Hunter interview, 1997.
53. Steven M. Spencer, 'Making a Heartbeat Behave,' *Saturday Evening Post* 234 (4 March 1961): 50; Bill Hakala and Nancy Skaran, *Bethesda: A Century of Caring, 1883–1983* (St. Paul, 1983), p. 83; Roth interview; Jeffrey, 'Pacing the Heart,' pp. 597–598; Mauston obituary, Minneapolis *Tribune*, 7 October 1966; Hunter interview, 1997.
54. Ibid. Oral tradition at Medtronic has it that one doctor called to complain about the awkwardness of implanting a pacemaker with such large handles: Peter Morawetz, personal communication to David Rhees. Amused condenscension toward the physicians is a subterranean theme in the medical device industry. For obvious reasons it cannot ever be expressed openly, but it shows up in anecdotes such as this. Similar legends circulate at all the manufacturing firms.
55. 'Toward Man's Full Life' (booklet, Minneapolis, Minn.: Medtronic, Inc., 1975), copy at Medtronic Information Resources Center, Fridley, Minn.
56. Jeffrey, 'Pacing the Heart,' discusses this dynamic in the 1960s and 1970s; for a broader application, see Fye, *American Cardiology*, esp. pp. 250–273, 301–306. For current practice guidelines see American College of Cardiology/American Heart Association Task Force on Practice Guidelines (Committee on Pacemaker Implantation), 'ACC/AHA Guidelines for Implantation of Cardiac Pacemakers and Antiarrhytymia Devices: Executive Summary,' *Circulation* 97 (7 April 1998): 1325–1335. On the changing understanding of chronic heart block in medical science, see Fye, 'A History of Cardiac Arrhythmias.'
57. Seymour Furman, 'Attempted Suicide,' *PACE* 3 (March-April 1980): 129.
58. According to Norman Roth, Warren Mauston 'liked to see the light blink. As long as the light was blinking, he was doing fine': Roth interview.
59. Bakken interview by Rhees, 10 January 1997. Bakken recalls that 'one night we had a lot of these [5800 pacemakers] going at the University of Minnesota, and the nurses started saying, well, sometimes at night these pacemakers quit. I couldn't figure that out—why should they quit at night? And so I finally took them in a darkroom and, sure enough, at times it appeared that the neon wasn't flashing. I didn't know whether it was an optical illusion or what it was.' Bakken tested one of the devices and found that 'sure enough, it did quit. The *output* didn't quit: it was running the same way; it was just that we had these adjusted so we would just barely trigger the neon light in an illuminated condition because we didn't want to wake [the children]. It was drawing half the current anyway, was going to flashing the bulb. We said, well, it's just at the threshold, and it took the ambient light to put it over the threshold so it would flash. We told them just to shine their flashlight on it. If it's flashing, everything is O.K. So that's all [it took].'
60. E.g., 'What it Means to have a Blue Baby,' *Woman's Home Companion* 79 (June 1952): 9+; 'How Blue Babies are Saved,' *Parents' Magazine* 31 (February 1956): 38+; 'Inside the Heart'; Steven M. Spencer, 'They Repair Damaged Hearts,' *Saturday Evening Post* 228 (7 April 1956): 32+; 'Ten Questions Parents Ask about Congenital Heart Defects,' *Today's Health* 38 (July 1960): 14+. Numerous newspaper articles reported on specific surgical cases; e.g., 'Electricity Aids in Heart Surgery,' New York *Times* (26 May 1956): 11; 'Hole in Heart Closed,' ibid. (15 August 1956): 31; 'Twenty Stitches in Heart Mend Girl, 14,' ibid. (17 September 1957): 29. On 6 May 1958, the DuMont television network presented a live broadcast of a heart operation (though not an open-heart procedure) on a three-year-old girl: New York *Times* (7 May 1958), p. 71. Bakken told the anecdote about his pastor in his interview with Rhees, 1 February 1998.
61. 'Inside the heart': 68.
62. A small electronics company in Pennsylvania called Atronics introduced a portable transistorized pulse generator a few months after the Medtronic 5800. But it was bulkier and enclosed in a dark-colored case, weighed two pounds, and lacked a red light that blinked reassuringly. Perhaps for these reasons, the Atronics generator is forgotten today. The device is pictured in 'Living Minute-to-Minute,' *Newsweek* 54 (6 July 1959): 54.
63. John Kasson, *Civilizing the Machine: Technology and Republican Values in America, 1776–1900* (New York, 1976), p. 185. Kasson's chapter 4 on 'the aesthetics of machinery' contains much that applies to the medical devices of the mid-twentieth century. In this context, the decision to change the color of the 5800 from black to white might be understood as an attempt to reinforce, through design, this reassuring quality. On medical devices as tokens of the future, see Nancy Knight, '"The

New Light: X-rays and Medical Futurism,' in *Imagining Tomorrow: History, Technology, and the American Future*, ed. Joseph J. Corn (Cambridge, MA, 1986), pp. 10–34. Kai N. Lee suggested to us the talismanic quality of the external pacemaker.

64. Spencer, 'Making a Heartbeat Behave.'
65. Samuel W. Hunter, interview by Kirk Jeffrey, 30 November 1989, NASPE Oral History Archive. A print of the five-minute film is housed at The Bakken Library and Museum; it was apparently made to explain cardiac pacing and medical electronics to potential investors.
66. Earl E. Bakken interview by Seymour Furman, 17 May 1996, NASPE Oral History Archive.
67. Ibid. Bob Wingrove, an early Medtronic engineer, sometimes assisted or substituted for Bakken on these visits to the VA Hospital. He said that he always had to remember to unplug the soldering iron before going to work—otherwise there would have been a direct connection from the electrical socket through the soldering iron and the lead to the patient's heart: Robert C. Wingrove, interview by Kirk Jeffrey, 28 April 1998, copy in interviewer's possession.
68. Both of the authors have attended this event, which is transmitted by satellite for employees at all Medtronic facilities around the world.
69. Bakken interview by Furman; italics added.
70. Ibid. Bakken meant that device manufacturers today often develop, test, and commercialize new devices outside the United States first because FDA rules unduly slow the process.
71. David Rhees and Ellen Kuhfeld, 'Notes on Examination of Earl Bakken's Prototype Transistorized External Pacemaker,' 5 September 1997. The device and the circuit diagram are housed at the Bakken Library and Museum in Minneapolis.
72. Kevin Olsen, *Minnesota Medical Device Manufacturing: An Analysis of Industry Growth, Specializations, and Location Factors* (St. Paul: Minnesota Department of Trade and Economic Opportunity, 1993). David J. Rhees and Eric Boyles, '"Medical Alley": The Origins of the Medical Device Industry in Minnesota,' a paper presented at the annual meeting of the Society for the History of Technology, London, August 1996, presents a full analysis of the origins of the device industry in Minnesota and the role of Medtronic.
73. Paul Weindling, 'From Infectious to Chronic Diseases: Changing Patterns of Sickness in the 19th and 20th Centuries', in *Medicine in Society: Historical Essays*, ed. Andrew Wear (Cambridge, Eng., 1992), pp. 303–316; M.V. Pauley, 'Taxation, Health Insurance, and Market Failure in the Medical Economy,' *Journal of Economic Literature* 25 (June 1986): 629–675; Rosemary Stevens, *American Medicine and the Public Interest: A History of Specialization* (2nd edn.; Berkeley and Los Angeles, 1998).
74. Terry Fiedler, 'Left to Their Own Devices,' *Minnesota Business Journal* (August 1982): 14–22; Marilyn L. Bach et al., 'A Medical Mecca in the Making,' *Business and Health* (November 1985): 43–46; Rhees and Boyles, '"Medical Alley."' See especially the special advertising supplement that Medical Alley published in *Time* magazine, 4 May 1987.
75. Kirk Jeffrey, *Machines in Our Hearts: The Cardiac Pacemaker, the Implantable Defibrillator, and American Health Care* (forthcoming), chapter 7. Medical Alley was born one year after the Health Care Financing Administration, the agency that administers Medicare, revamped its system of reimbursing hospitals (and later physicians) for treating Medicare patients. The changes encouraged care providers to economize on the use of high-tech medical equipment and devices in diagnosis and treatment.
76. Todd Nissen, 'The Future of the Medical Device Industry,' *Corporate Report Minnesota*, December 1993, sec. 1, p. 54; L.R. Pilot, 'The Safe Medical Devices Act of 1990: The Challenge Ahead,' *Medical Device & Diagnostic Industry* 13 (January 1991): <hhtp://ww.devicelink.com/mddi/archive/>; Alan H. Magazine and Michie I. Hunt, 'The U.S. Medical Device Industry Moves Offshore,' ibid. 17 (August 1995): 72–79.
77. James Thornton, 'R&D with the U,' *Corporate Report Minnesota*, March 1987, 45–48.
78. A fundraising effort for the Biomedical Engineering Institute (or Center, as it then was called) had begun in 1987 but had been aborted, apparently because of a change in university leadership and the onset of a national recession.
79. Thornton, 'R&D with the U,' 46; 'Medical Alley: Where Medical Minds Develop New Technologies,' *Time* (4 May 1987), special advertising section.
80. Michael P. Moore, 'The Genesis of Minnesota's Medical Alley,' *University of Minnesota Medical Bulletin* (Winter 1992): 7–13.
81. Information supplied by Ronald Hagenson..

82. The Mission Statement speaks of contributing to human welfare through biomedical engineering, focusing resources in areas where the company can 'make unique and worthy contributions,' striving for product reliability and quality, making a fair profit, recognizing the worth of employees, and maintaining good company citizenship. It is reprinted in full, in ten languages, in the 1998 Medtronic annual report. Bakken discussed the origins of the Mission Statement in a talk at the Minnesota Center for Corporate Responsibility, Graduate School of Business, University of St. Thomas (Minneapolis: University of St. Thomas, typescript, May 1991; copy courtesy of Thomas E. Holloran). Earl Bakken has been strongly interested in origins and continuities. When the company was still very small, he began collecting books and scientific apparatus illustrating the history of therapeutic electrostimulation; his collection forms the core of The Bakken Library and Museum in Minneapolis.
83. Thomas E. Holloran, interview by Kirk Jeffrey, 4 March 1997; Daniel R. Denison, *Corporate Culture and Organizational Effectiveness* (New York, 1990), pp. 98–100. See also James C. Collins and Jerry I. Porras, *Built to Last: Successful Habits of Visionary Companies* (New York, 1994).
84. Except one—the Xytron (1973), the only Medtronic pulse generator on which the company has had to issue product advisories or 'recalls.'
85. Thomas M. Burton, 'Pacemaker-style Devices Mend the Nervous System,' *Wall Street Journal* (Interactive Edition) (6 January 1998); Medtronic annual report, 1998; Bakken interview by Rhees, 28 August 1997, 'Pioneers of the Medical Device Industry' Oral History Series, Minnesota Historical Society.
86. Rhees interview with Bakken and Lillehei, 9 September 1997. C. Walton Lillehei died in July 1999.
87. Earl E. Bakken, *One Man's Full Life* (Minneapolis: Medtronic, Inc., 1999). On value-driven corporations see Collins and Porras, *Built to Last.*

Ross Bassett

When is a Microprocessor not a Microprocessor? The Industrial Construction of Semiconductor Innovation[1]

In the early 1990s an integrated circuit first made in 1969 and thus antedating by two years the chip typically seen as the first microprocessor (Intel's 4004), became a microprocessor for the first time. The stimulus for this piece of industrial alchemy was a patent fight. A microprocessor patent had been issued to Texas Instruments, and companies faced with patent infringement lawsuits were looking for prior art with which to challenge it.[2] This old integrated circuit, but new microprocessor, was the AL1, designed by Lee Boysel and used in computers built by his start-up, Four-Phase Systems, established in 1968. In its 1990s reincarnation a demonstration system was built showing that the AL1 could have operated according to the classic microprocessor model, with ROM (Read Only Memory), RAM (Random Access Memory), and I/O (Input/Output) forming a basic computer. The operative words here are could have, for it was never used in that configuration during its normal lifetime. Instead it was used as one-third of a 24-bit CPU (Central Processing Unit) for a series of computers built by Four-Phase.[3]

Examining the AL1 through the lenses of the history of technology and business history puts Intel's microprocessor work into a different perspective. The differences between Four-Phase's and Intel's work were industrially constructed; they owed much to the different industries each saw itself in.[4] While putting a substantial part of a central processing unit on a chip was not a discrete invention for Four-Phase or the computer industry, it was in the semiconductor industry. Although the AL1 was in many ways technically superior to Intel's first generation microprocessors, its location in the computer industry led to a different, and ultimately truncated development trajectory. Flexibility was the hallmark of Intel's microprocessor, with Intel and its customers finding countless applications for it, while the industrially constructed rigidity of the AL1 limited its applications.

The story of the AL1 and Four-Phase Systems provides a case study of a start-up in Silicon Valley's adolescent period. Four-Phase owed its origins to a visionary engineer working for an inattentive wealthy company willing to fund work in a new area, but unable (or unwilling) to manage it so that it would benefit the company. Four-Phase is also important as a representative of the successful Silicon Valley firms that do not achieve the level of public visibility of an Intel or Apple. Looking at both Four-Phase and Intel provides a sense of the diversity of Silicon

Figure 1. Lee Boysel (right) with Cloyd Marvin, another former Fairchild engineer who moved to Four-Phase Systems, looking at the layout of Four-Phase's AL1 integrated circuit. Courtesy Lee Boysel.

Valley start-ups during this period, a diversity that increased the chance that some would succeed.[5]

Lee Boysel, MOS Maverick

The essential background to both the AL1 and Intel's 4004 is the MOS (Metal-Oxide-Semiconductor) technology and the MOS work done at Fairchild Semiconductor in the 1960s. The MOS transistor came onto the agenda of the semiconductor industry in the early 1960s, primarily because it was simpler to build than the dominant bipolar transistor, allowing one to put many more transistors on a chip. However its drawbacks included a dependence on the characteristics of the surface of the silicon, which could not be well controlled, and a much slower speed than the bipolar transistor. Advocates of MOS technology searched for an application which would take advantage of its characteristics.[6]

Fairchild Semiconductor, the semiconductor industry's most dynamic firm in the 1960s, supported two different and competing MOS programs. The primary one was at Fairchild's research and development laboratory in Palo Alto, California under the direction of Gordon Moore. A major part of the R&D program was devoted to understanding the chemistry and physics of MOS structures and developing methods of fabricating them. The group was by and large made up of chemists, physicists, chemical engineers, and electrical engineers concerned with the physical and chemical processes involved in making semiconductors. (They had almost no background in computing.) The nucleus of this effort left to form Intel.[7] The other group, almost a bootleg operation, was in applications, a few miles down the road in Mountain View, but a world away in its approach to the technology. This group consisted of electrical engineers who were interested in using MOS technology to design complex systems on a single chip. While they were not strong in semiconductor processing, they understood computers and circuits. They were led by Lee Boysel, a highly creative and maverick MOS devotee.[8]

Lee Boysel's MOS work started with an epiphany two years before he joined Fairchild. In 1964 he was an electrical engineer and electronics enthusiast just out of the University of Michigan, working at Douglas Aircraft in Santa Monica, California. At that time Frank Wanlass, one of the first proponents of MOS technology, visited and showed him that a twenty-bit shift register could be built on a single MOS chip, something that would have taken many circuit boards using individual transistors.[9] Boysel became a believer in the possibilities of MOS technology and from that point worked on it exclusively. He designed MOS circuits at Douglas and then moved to IBM in Huntsville, Alabama, where he designed MOS circuits for use in space applications.[10]

Boysel was highly individualistic, unwilling to sacrifice his own personal goals for the good of the larger organization. The key characteristics of his working style were a preference for hands-on work over analysis, a willingness to work extremely long hours, and an impatience that led him to circumvent formal bureaucratic channels. He had a home laboratory equipped with government surplus equipment he had bought while he was at Douglas. On the evenings and weekends, he would frequently carry his circuit design projects back to his home lab. While this might suggest someone who was doing the company's work on his own time, in reality the opposite was much more nearly true. At both Douglas and IBM he had carried his MOS efforts far beyond his mandate or the companies' interests; this pattern would continue when he joined Fairchild Semiconductor in 1966.

Boysel's experience at Fairchild shows how a person in a marginal position at a large corporation could use that position to accomplish his goals even when they were not congruent with the corporation's. At the time he joined Fairchild, he benefitted from the uncertainty surrounding the unproven MOS technology. No one knew which, if any, applications would be successful. MOS technology at the time was a minuscule portion of Fairchild's business, and managers concerned themselves more with the dominant bipolar technology. Boysel was in a marginal position at Fairchild and took full advantage of it. While Fairchild managers might have a justifiable tentativeness in their approach to the new technology, Boysel knew exactly what he wanted to do and charged ahead. He designed a 256-bit ROM before his management knew anything

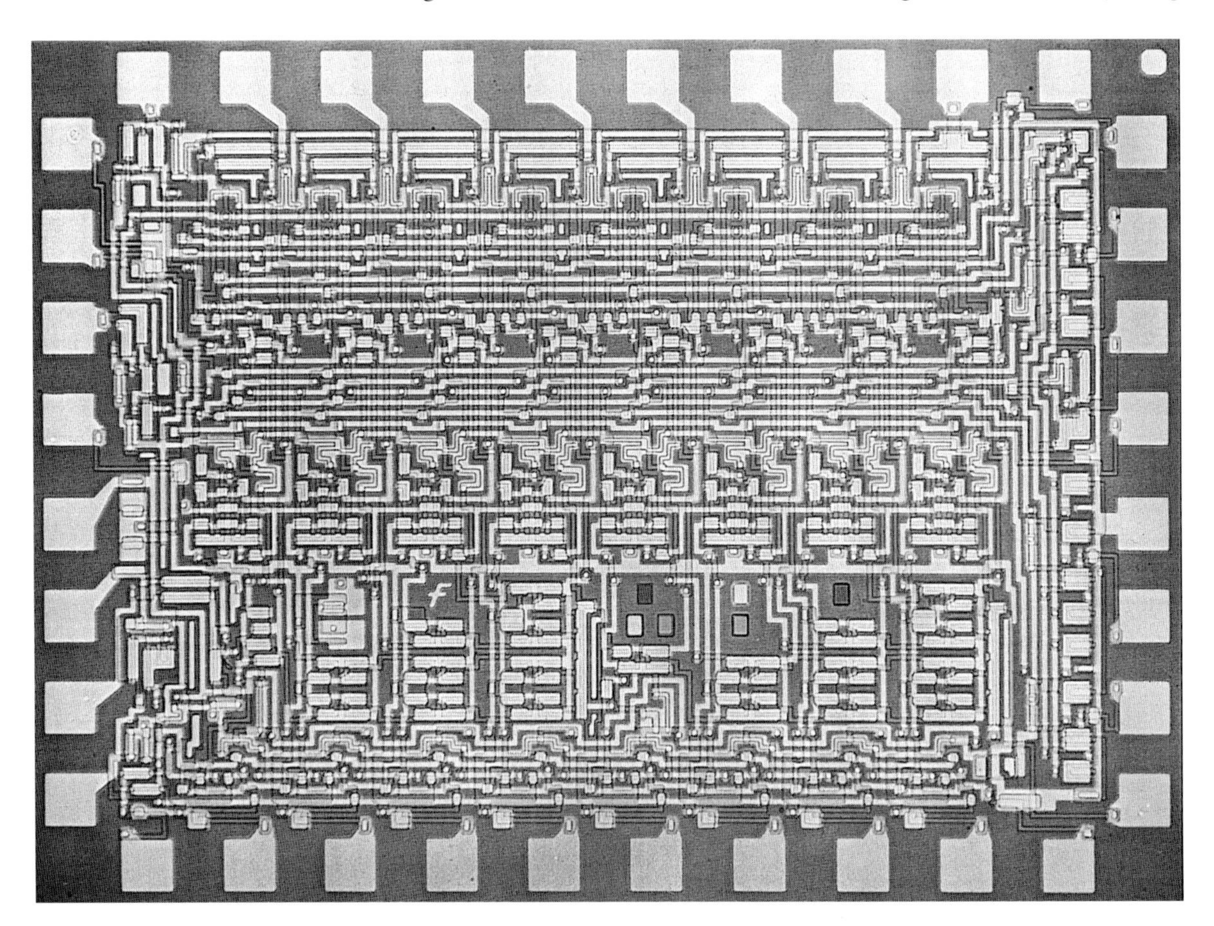

Figure 2. Photograph of eight-bit arithmetic unit integrated circuit designed by Lee Boysel at Fairchild circa 1967. This photograph comes from a Fairchild publicity booklet of the late 1970s or early 1980s containing pictures of some of the most important integrated circuits Fairchild had designed. The integrated circuit was called 'the first standard MOS product for data processing applications.' No mention was made of its lack of success in the market or of Boysel's role in designing the part. The size of the chip is 110 × 80 mils. Source: Fairchild Camera and Instrument, A Solid State of Progress (n.p.: Fairchild Camera and Instrument, n.d.).

about it. He followed this with an eight-bit adder in MOS. Neither part was commercially successful.[11]

In 1967 Boysel wrote a manifesto putting his work into perspective and revealing his plans. In a single sheet of paper consisting of a computer block diagram and a third of a page of text, Boysel showed that an entire computer (including memory) could be made out of MOS technology. At this time no computer in production had been built using semiconductor memory, MOS or bipolar. Proposing a computer built entirely from bipolar technology would have exhibited forward thinking; using MOS technology was radical.[12]

The bulk of the computer would be made from six different MOS part numbers. He described the CPU element as a '4 bit wide slice with all op code decoding and branch instruction built-in.' The ROM chip (at IBM he had studied the System/360 architecture and gotten the idea of microprogramming using ROM) would contain either four to eight thousand bits, while the RAM chip would contain either 512 or 1024 bits. Perhaps Boysel's most extraordinary claim was that one could build a bona fide computer out of MOS technology. At the time most people thought MOS technology was best suited for calculators; Boysel claimed that with proper design, it could be used to build mainframe computers.[13]

Although the manifesto included a business analysis, purporting to show why it made sense for Fairchild, Boysel's vision was a personal one, not closely tied into considerations of Fairchild strategy. Boysel's vision was one that ultimately would not be satisfying to either Fairchild or its customers. Were customers to adopt Boysel's approach, it would take away dramatically from Fairchild's existing bipolar business and require a radical reorganization of the company. In any case, customers had ample reason to be hostile to Boysel's plans. The use of small scale integration (that is a few logic circuits on a chip) allowed computer makers to design and implement a computer in the way they believed optimal. Boysel was proposing to reparse the task so that his group encroached on the responsibilities of computer system designers. The jurisdictional question was compounded by the fact that Boysel's plan threatened to take away one of the computer makers' main sources of added value, their design of the computer. In any case Boysel did not have Fairchild's interests in mind: he was intending to start his own computer company and his manifesto described the approach he would use. Fairchild was unwittingly providing the early development funding for this start-up.[14]

Although none of Boysel's designs was a business success, Fairchild, flush with money and not fully aware of his motives, authorized him to hire others to form a MOS design group. Fairchild management appears to have valued Boysel because he gave the company a credible presence in MOS. (While the long-term outlook for MOS was uncertain, a few start-ups had gathered a lot of attention in the trade press for their work on MOS.) Boysel and his group designed enough MOS parts to fill a catalog, although none of them were sold in any volumes. He wrote several articles in the trade press describing his MOS parts, even though he had designed them not with Fairchild's need in mind, but the requirements of his own computer.[15]

Given the marginal status of MOS technology at Fairchild, and Boysel's view of himself as an insurgent, he was successfully able to create an environment where those he hired saw their primary allegiance to him, rather than to Fairchild Semiconductor. By mid-1968, Boysel and his group had designed the key parts necessary to build a computer from MOS technology. With his initial development work done, he left to start his own company in October 1968.[16]

Boysel's company started as a partnership, with two other engineers from his group at Fairchild joined by several outsiders. They worked in a rented dentist's office, using equipment Boysel had in his home laboratory. No one was paid while Boysel sought long term funding. Four-Phase Systems was incorporated in February 1969, based on $2 million of long-term notes. A large portion of the funding came from Corning Glass Works, which also owned Signetics, a major semiconductor manufacturer. At this point, other members of his Fairchild group joined him. The name Four-Phase hinted at Boysel's technological enthusiasm, for it was an abstruse type of MOS circuitry.[17]

Boysel planned to build the computer he had described in his September 1967 MOS design manifesto. The computer would use MOS technology throughout. Through the use of four-phase circuitry, which made possible the use of very small MOS transistors, Boysel was able to build much denser chips than he had envisioned earlier. The computer would be equivalent in power to a medium-sized IBM System/360. Boysel and his company sought a niche for their system, finally settling on using it as a terminal controller, specifically a plug-compatible replacement for IBM systems. This provided Boysel access to an established market and lowered the risk for potential customers, for if his system proved unsatisfactory, they could quickly replace it with an IBM system.[18]

Since Four-Phase was built on Boysel's integrated circuit designs but a computer start-up could not justify the costs of a dedicated semiconductor fabrication facility, Boysel made arrangements to assure the company access to integrated circuit manufacturing capacity. At the time Boysel left, a Fairchild colleague who worked in MOS manufacturing also left to start Cartesian, a company which would process MOS wafers for Four-Phase and other companies who were able to design their own circuits. Cartesian implemented the MOS process that had been used at Fairchild manufacturing, and that Boysel and his group had used in all their previous designs. Without this continuity in the manufacturing process, Four-Phase would have had to substantially modify its circuit designs, extra work that would be costly in time and money for a start-up. The two firms formed a dyad, with Cartesian's creation essential to Four-Phase, and Boysel arranging financing for it along with his own company.[19]

By the spring of 1970, Four-Phase had an engineering-level system operating, and it publicly introduced its system that fall. By June of 1971, Four-Phase systems were in operation at Eastern Airlines, United Airlines, Bankers Trust, and McDonnell-Douglas. One user asserted the cost of the Four-Phase system was roughly half of an equivalent IBM system. All users interviewed in an article in *Computerworld* gave very positive reports on their experiences with the system. By March 1973, Four-Phase had shipped 347 systems, with 3,929 terminals to 131 different customers.[20]

The AL1

The heart of the system was the AL1 chip, which Boysel had conceived of and designed while still at Fairchild. The AL1 contained an eight-bit arithmetic unit and eight eight-bit registers (including the program counter). It was an extremely complex chip, with over a thousand logic gates in an area of 130 × 120 mils. (To give a point of comparison with later integrated circuits, this was roughly the same number of logic circuits in Intel's 8008 in an area the size of Intel's 4004.)

One should point out the obvious: there was no such thing as a microcomputer or a computer on a chip at this time. (At the time, the term microprocessor had a different definition, meaning a microprogrammed processor with RAM and ROM.) Having one's integrated circuit recognized as a computer on a chip depended not on meeting some fixed set of criteria, but on making a claim to the title and subsequently convincing the semiconductor and computing communities of the validity of that claim. Boysel and Four-Phase might have called the AL1 a computer on a chip and one can imagine a scenario where this claim would have later engendered some debate with Intel and others about what constituted a computer on a chip. Boysel made no claims for his chip and there was no such discussion. And in fact, Boysel and Four-Phase seem to have had a hard time coming up with a descriptive name for the AL1. In an April 1970 article in *Computer Design*, Boysel and one of his colleagues alternately called the AL1 'the main LSI block of a low-cost fourth-generation commercial computer system,' an 'eight-bit computer slice,' and an 'arithmetic logic block.'[21]

Although the AL1 was an outstanding piece of engineering work, essential to the success of his overall computer, Boysel did not see it as an innovation in and of itself. Boysel had done two previous designs of adders or arithmetic units at Fairchild and the AL1 represented simply a continuation of that work; it was a change in degree not a change in kind. The computer system that Four-Phase had developed had required the design and fabrication of a number of complex and innovative inte-

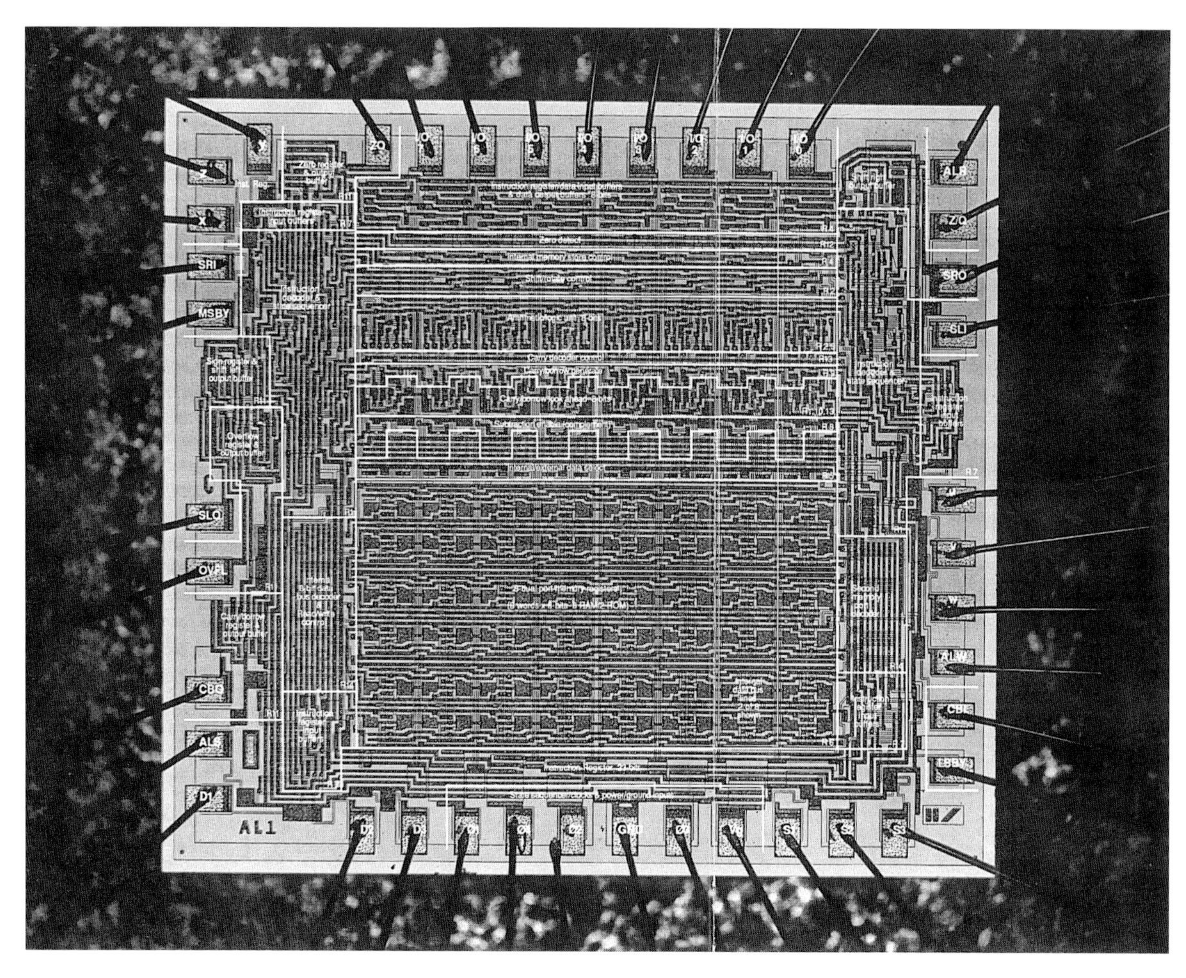

Figure 3. Photograph of AL1. As part of the AL1's 1990s reincarnation as a microprocessor, this photograph has overlays identifying the chip's different functional elements, a practice common in microprocessors. Notice that the AL1 has forty input/outputs, the pads on the perimeter of the chip. The size of the chip is 130 × 1210 mils. Courtesy Lee Boysel.

grated circuits. Among these, the AL1 was not uniquely important or innovative—each was needed to build the system.

To Boysel the AL1 was neither a computer on a chip nor a processor on a chip. First of all, because Four-Phase's computer was 24-bits wide, three AL1's were required per system. Furthermore in Four-Phases's nomenclature the Central Processing Unit (CPU) circuit board was made up of three AL1 chips, three chips called Random Logic (RL), and three ROMs. The CPU included the control store (ROM). The AL1 was not a microprocessor under the existing definition of that term.

In 1971 and 1972 respectively, Intel introduced the 4004 and 8008 integrated circuits which are now considered the first microprocessors. [22] By metaphorically putting the AL1 alongside them, we can get some idea of how Four-Phase's position in the computer industry shaped the AL1 and its development. Intel trumpeted the announcement of the 4004 as 'a new era of integrated electronics' and a 'microprogrammable computer on a chip,' for the obvious reason that the more this thing caught the users' eye, the more likely they were to buy it. (The 4004 was patently

Figure 4. Photograph of the 4004, Intel's original 4-bit microprocessor chip. Notice the 16 input/output pads around the periphery of the chip. The chip size is 170 × 125 mils. Courtesy Intel Corporation.

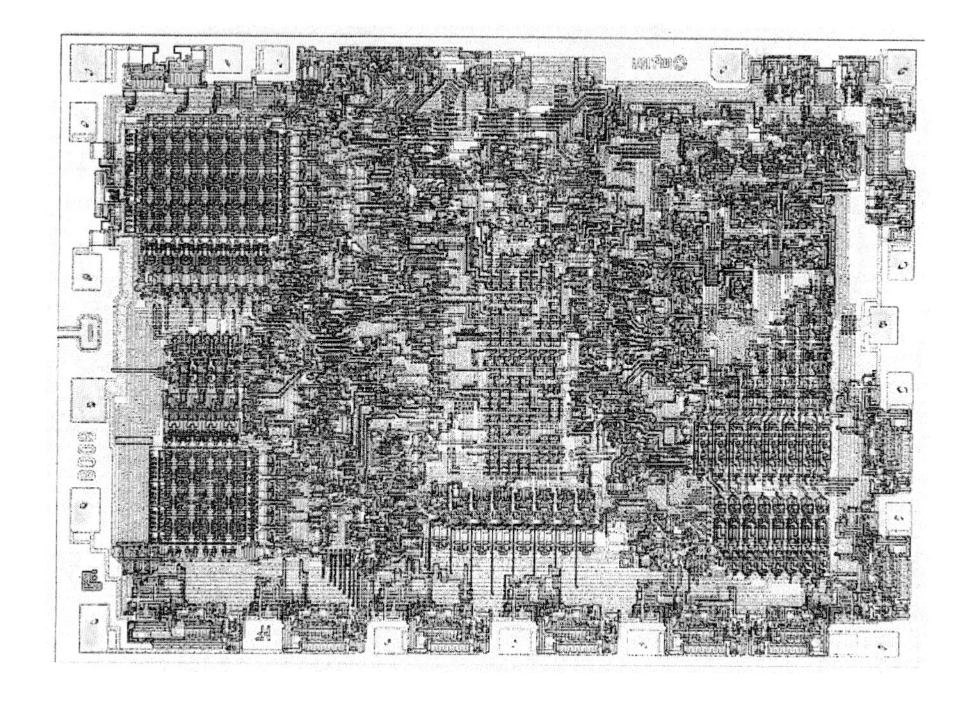

Figure 5. Photomicrograph of the 8008, Intel's 8-bit microprocessor chip. Notice the 18 input/out pads around the chip's periphery. The chip size is 170 × 125 mils. Courtesy Intel Corporation.

not a computer on a chip; even calling it a CPU on a chip represented a diminution of that term.) The AL1 received no such publicity. What mattered to Boysel, as an entrant into the computer business, were the characteristics of the overall system, such as performance or cost. Unless it offered some such system benefit a customer could readily understand, he had no reason to claim his AL1 represented something new. Boysel's nonchalant attitude toward putting a computer on a chip can be seen in his statements made for an article published in *Electronic Design* in February 1970. At this time what would become the 4004 and 8008 microprocessors from Intel existed only as block diagrams, but Four-Phase had a working computer system built around the AL1 chip. Boysel asserted in the article:

> The computer on a chip is no big deal. It's almost here now. We're down to nine chips and we're not even pushing the state of the art. I've no doubt that the whole computer will be on one chip within five years.[23]

In fact, being in the computer business gave Boysel reason to be quiet about the AL1. While there was no reason to believe that a sale would

Figure 6. Early ad for Four-Phase Systems Model IV/70, circa 1970. The following page states 'then: You'd have System IV/70—the data base access system of the 70s.' Note that the advertisement only mentions the IV/70 having 'the whole 75,000 component CPU on one card.' The top circuit card above '2' is the CPU card, with a random logic chip and ROM chips. The lower circuit card is the memory card, which used Four-Phase's 1K RAM chips. Courtesy, Lee Boysel.

hinge on exactly how much of the central processing unit Four-Phase had put on one integrated circuit, the AL1 was one of Four-Phase's most valuable pieces of intellectual property and were other companies to copy it, they could conceivably offer a competitive product. Four-Phase refused to sell the AL1 to a manufacturer of computer terminals, believing that such a sale would jeopardize its computer business, which it considered more profitable than the semiconductor business.[24]

Even when Four-Phase did want to talk about the advanced semiconductor technology in its systems as a means of promoting sales, there were other things besides the AL1 to point to. Four-Phase produced one of the first computer systems equipped with MOS semiconductor memory in place of magnetic cores. Semiconductor memories had been the subject of much discussion and work in both the semiconductor and computer industries, which would make it easier for customers to understand the significance of Four-Phase's achievement. An article by Boysel and two colleagues on Four-Phase's semiconductor memory was the cover story in *Electronics*, the leading trade journal in the industry.[25]

Four Phase's position in the computer industry also shaped the way the AL1 was packaged. A traditional constraint in the packaging of integrated circuits was in the number of inputs and output pins on the package. In the late 1960s low-cost dual-inline-packages (DIPs) were available with 16 or 18 pins. For an eight-bit processor, the use of such packages would mean that there were more input and output signals than there were pins on the package. One solution to this problem was to have some signal lines share pins, which slowed down the system. Another alternative was to use a very expensive 40-pin package. Costs for such a package were so high (around five dollars each), that they represented a greater cost than making the integrated circuit itself.[26]

Intel put the 8008 in an 18-pin DIP and multiplexed the signals. It had used these packages for its memory chips and they were cheap. Intel, as would be expected of a semiconductor producer, was highly resistant to using an expensive purchased package that would put a majority of the cost of the component outside of its control. The AL1 was put in a 40-pin DIP. For a computer systems company, building a computer that would have a purchase price of ten thousand dollars or more, the few extra dollars paid for high pin count packages were more than made up for in increased performance. Intel's next generation microprocessor, the 8080, did come in a 40-pin DIP, after customers complained that the 18-pin DIP hobbled the 8008's performance, and after package prices came down.[27]

For all Intel's talk about a computer on a chip, the 8008 required many ancillary chips to convert voltage levels so it could interface with other subsystems, such as memory. In spite of what Intel advertising might say, the 8008 was not part of an integrally designed system, it was an opportunistically designed part. No such interface circuits were required with the AL1, for the whole system was designed together. Again, in its second generation microprocessor, Intel fixed this after receiving complaints from customers.[28]

One of the ways the AL1 got its speed was through the use of four-phase logic, a very sophisticated type of circuitry, which had used four clock signals applied to the chip in a very close relationship. This required special clock circuits off the chip and a very knowledgeable user, who could control the parameters of the entire system, such as wiring lengths, to assure the system would work. Intel did not use four-phase logic, and the 4004 and 8008 could tolerate a much less sophisticated user.[29]

The Microprocessor: an Innovation in the Semiconductor Industry

The idea here is not to substitute the name of Lee Boysel for Ted Hoff as the inventor of the microprocessor, but to suggest that the AL1 can provide a different perspective on the invention of the microprocessor. Eric von Hippel has written that an innovation often occurs in the industry that stands to benefit the most from it. Here we have something analogous with respect to what constitutes an invention. [30] Looking at Four-Phase and Intel, similar integrated circuits were being made in both firms, one in the computer industry and one in the semiconductor industry, but only the semiconductor company called it an invention. The economics of the semiconductor industry gave what we call the microprocessor a completely different meaning than it had in the computer industry.

Perhaps not unrelatedly, a number of Hoff's contemporaries who were outside the semiconductor industry and familiar with computers have not been much impressed with the microprocessor as an invention. Carver Mead, a professor of electrical engineering at Caltech working on applications of digital electronics, called it a 'no-brainer.' Researchers from IBM claimed that they did not consider the Intel work an invention at all. In a manual written in 1968, before Intel had done anything, one IBM researcher wrote 'hopefully the day of the "computer on a slice" is nearly dawning.' [31] It was, but it did not dawn at IBM; it rose in the west first. The IBMers knew something like a microprocessor could be made, but they had no incentive to do it. It made little difference to a computer company whether the central processing unit function was put on one integrated circuit or on several. For something like a computer-on-a-chip to be significant to a computer company, there would have to be additional innovations which would make new computer systems possible.

To Intel however, the microprocessor was an innovation that was a solution to a real problem. Gordon Moore and Robert Noyce had founded Intel as a company that would concentrate on standard Large Scale Integration (LSI) products. The possibility of putting many transistors on an integrated circuit raised what became known as the part-number-problem. When each integrated circuit contained only one or two logic circuits, it was possible to build up a large digital system from these primitives using only a handful of different chip types. But when each chip contained upwards of a hundred logic circuits as envisioned in LSI, each chip became confined to a specific application in a specific system. Since the number of circuits in a system was still large compared to the one hundred on a chip, a system built using LSI would be made up of many chips, and also many different chip part numbers. Across different systems, there would be very little commonality, so that part numbers could not be shared. This explosion of part numbers, where the semiconductor plant would have to make small quantities of numerous chip designs, threatened to undermine any economic advantage that accrued by moving to higher levels of integration.[32]

In the late 1960s, three approaches to building LSI chips existed: computer-aided-design, custom, and standard. Under computer-aided-design, the semiconductor company would develop an infrastructure that customers would use to do the design work themselves. Automation would allow a semiconductor company economically to produce a large number of different part numbers in small volumes. Moore had had a large computer-aided-design program underway at Fairchild, but he developed doubts about it shortly before he left to form Intel. In its early days, Intel made no efforts in this area.[33]

The custom approach involved dealing with the part-number problem on a case by case basis. A semiconductor company would contract to design a specific part for a specific company. For such an approach to work, the parts had to have high volumes. The problem with this approach was that it required the semiconductor company to have a large number of designers, and potential customers could be very fickle. After the semiconductor company had put a great deal of effort into the design of a part, the customer's requirements might change and the order vanish.[34]

The last approach was standard parts. The semiconductor company would design parts which it believed had general applicability and offer them for sale to all comers. This approach could lead to large production volumes for every part, and would require few circuit designers. One disadvantage of this approach was that it could expose the company to competition as other firms moved in to make a popular part.

As they started Intel, Moore and Noyce believed that standard parts were the only economically viable way to proceed with LSI. At the 1966 Fall Joint Computer Conference session considering large scale integration Robert Noyce—still at Fairchild—spoke of the cost advantages of standard parts, and asserted that how far the industry accepted standardization 'may well determine whether or not large arrays [i.e., LSI] are used in significant quantity in the future.' While Noyce noted that at the time the only standard LSI arrays in existence were memories and shift registers, he also prophesied that '[t]he appearance of more standard arrays seems inevitable.' In the spring of 1970, before Intel's first microprocessor had been implemented, Robert Noyce spoke at the IEEE annual convention on trends in silicon technology. He noted Moore's law (the doubling of the number of components on a chip every year), but asserted that one of the major potential limitations of silicon integrated circuits was an economic one. He claimed that 'finding high volume "universal" high complexity circuits will be difficult,' and 'failing in this quest, the fabrication costs become secondary to the costs of design and tooling,' resulting in a lack of motivation to produce more complex integrated circuits.[35]

Intel was started by Moore and Noyce as a company to make standard LSI products, and its early work was in memories because they were the ones then in existence.[36] But Noyce and Moore were predisposed to embrace the idea of a processor on a chip, because of its potential as a standard part. Much of the writing on the microprocessor has focused on Ted Hoff's genius, or how credit should be apportioned between Hoff and the other Intel engineers, while ignoring the larger issue of corporate strategy.[37]

The microprocessor gave Intel access to the largest market for digital integrated circuits, logic, on its terms. Intel's core strength was in semiconductor processing technology and it was founded on the premise that it could gain a competitive advantage through the development of a new process that would allow for the fabrication of very complex integrated circuits. Intel could thereby stay out of the market for simple small scale integration circuits, which were commodity parts subject to vicious price wars. But prior to the microprocessor, Intel's commitment to LSI excluded it from the market for digital logic, because no standard LSI logic parts existed. Most digital logic applications required much less than a full general purpose computer and were made up of many small scale integration Transistor-Transistor Logic (TTL) integrated circuits. With the introduction of the 4004 and 8008, Intel proposed that its customers reparse their systems to replace many cheap integrated circuits (made by someone else) with a few high priced integrated circuits (made by Intel). In an August 1972 ad Intel claimed that its 4004 could typically replace 90 TTL circuits, while the 8008 could replace 125.[38]

The important fact about Intel's microprocessor was not that it was a computer-on-a-chip, but that it was generalizable and could aggregate demand. Based on this, the trajectories of Boysel's AL1 and Intel's microprocessors greatly diverged. Intel's microprocessors could be used for a wide variety of things. Intel and its consultants came up with a notebook full of possible applications. The applications of the microprocessor were not at all limited by what Intel engineers could conceive. A frequent occurrence at Intel's microprocessor seminars was for a participant to come up afterwards and present an Intel engineer with a new way of using the chip. The typical response was, 'I hadn't thought of that, but yeah it would work.'[39]

The most dramatic example of how the microprocessor was not limited to uses Intel supported comes with the personal computer. Gordon Moore conceived of a personal computer as one possible application early in the history of the microprocessor, but he ultimately rejected it as a bad idea. The only use he could think of for it was as a place for housewives to keep recipes.[40] Of course that did not mean that no one could attempt to build a personal computer with an Intel microprocessor, and others did.[41]

On the other hand, while the AL1 had the architecture of a general purpose computer, making it capable of being used in a variety of applications, it was under the complete control of Four-Phase and it could only be used as one third of the Arithmetic Logic Unit of a Four-Phase system unless Four-Phase decided otherwise. It is probably fair to say that few of Four-Phase's customers even knew (or cared) that such a thing as an AL1 even existed. While they could come up with new uses for an entire Four-Phase system, they could not come up with new applications for the AL1. Any innovations in how the AL1 would be used had to come from Four-Phase itself. And while Boysel was an extremely creative person, he had his hands full running the company.

Four-Phase's position in the computer industry further constrained the proliferation of the AL1 chip. In the early 1970s, while the semiconductor industry was not capital intensive, the computer industry was, due to the way medium to large-sized computers were acquired. Following a pattern that had been set in the pre-computer office equipment era, most computers were leased, not bought. This meant that every computer that Four-Phase made was a capital item, which would only gradually pay for its costs over the course of its lease. Four-Phase had to raise funds, either debt or equity, to pay for every computer it made. Leasing acted as a rein on the growth of a start-up, for the capital requirements would strangle the company if it were to grow too quickly. By 1974, the company's first full year of profitability, Boysel had had to raise twenty-seven million dollars to keep it going.[42]

Leasing made Boysel, at heart a technological radical, more conservative. Each new model that Boysel introduced threatened his installed lease base. Four-Phase's conservatism can be seen in its use of semiconductor technology. Four-Phase was able to quickly produce enough chips to meet its requirements into the foreseeable future. Production was halted and the chips were stored for later use. Such action would have been inconceivable in the semiconductor industry, where a part's value only went down over time. But at Four-Phase the value of the chips was related to the lease price Four-Phase could get for its systems, which remained relatively constant over time. While the semiconductor industry had a highly elastic demand for its chips, Four-Phase faced an inelastic demand; even if its chips could be produced for nothing, the intricacies of lease financing would determine how many systems Four-Phase would build.

Although it never became the size of IBM, or even Digital Equipment, Four-Phase achieved a substantial level of success with its approach. Its systems were widely used by hospitals to handle billing as well as by government agencies for data entry. Although Four-Phase Annual Reports made constant reference to the continual need for capital (i.e. bank loans) required in the computer leasing business, Four-Phase stayed in the good graces of the banks and financial markets. In 1979 the firm had revenues of $178 million and net income of $16 million. In 1982, in the face of increasingly aggressive competition from IBM, Four-Phase was sold to Motorola in an exchange of stock valued at $253 million.[43]

Conclusion

Nathan Rosenberg's classic article on the machine tool industry reminds us that innovations may be more likely to enter the economy through a particular door (or industry). Here that door was the semiconductor industry. For although they looked similar at the block diagram level, the AL1 was an answer to a specific problem; Intel's microprocessor could respond to a range of generalizable problems.[44]

Lee Boysel's comment in 1970 that the 'computer on a chip was no big deal' was only half right. Many people had seen it coming, and for a computer company that made its own semiconductors, it was of little moment whether the central processing unit was on one chip or two or three. But for a semiconductor company such as Intel, what it called the computer on a chip had great import. It offered a way for Intel to get into markets previously denied it, and to bring electronics into a wide new range of areas. Gordon Moore stated that it allowed Intel to 'make a single microprocessor chip and sell it for several thousand applications.' In 1975 Robert Noyce was calling Intel 'the world's largest computer manufacturer.'[45] It would take years before that was manifest to the rest of the world.

Notes

1. A draft of this article was presented at the Society for the History of Technology's 1997 meeting in Pasadena, California. I would like to thank Steven Usselman, Barney Finn, Michael S. Mahoney and an anonymous reviewer for helpful comments. This work was supported by a grant from the National Science Foundation and a SDF Project Fellowship from the Stanford University Libraries.
2. For Texas Instrument's patent suits in the 1990s, see *New York Times* (11 September 1990), p. D4; (9 March 1993), p. D18; (28 April 1993), p. D4. The suits were settled out of court. For a discussion of the Texas Instruments microprocessor patent, see Michael S. Malone, *The Microprocessor: A Biography* (New York, 1995), pp. 13–14, 130–132. In an example of the shifting connections common in Silicon Valley, the work on the AL1 as microprocessor was coordinated by an intellectual property firm whose chief technical officer was Ted Hoff, the putative inventor of the microprocessor.

3. Lee Boysel, 'Court Room Demonstration System, 1969 AL1 Microprocessor,' (3 April 1995) (in author's possession); Lee Boysel, 'AL1 Microprocessor Demonstration Model,' (27 February 1993) (in author's possession).
4. Looking at how technologies are shaped differently in different industries is compatible with Pinch and Bijker's social construction of technology and may be a way of synthesizing business history and the history of technology. Trevor J. Pinch and Wiebe E. Bijker, 'The Social Construction of Facts and Artifacts: Or How the Sociology of Science and the Sociology of Technology Might Benefit Each Other,' in Wiebe E. Bijker, Thomas P. Hughes, and Trevor J. Pinch (eds.), *The Social Construction of Technological Systems*, (Cambridge, MA, 1987), pp. 17–50. For a summary of the state of the relationship between business history and the history of technology, see David A. Hounshell, 'Hughesian History of Technology and Chandlerian Business History: Parallels, Departures and Critics,' *History and Technology* 12 (1995), 205–24. For a study examining how the determination of what constitutes an invention is socially constructed, see Carolyn C. Cooper, *Shaping Invention: Thomas Blanchard's Machinery and Patent Management in Nineteenth-Century America* (New York, 1991).
5. For a study of Silicon Valley that takes an ecological approach, see AnnaLee Saxenian, *Regional Advantage: Culture and Competition in Silicon Valley and Route 128* (Cambridge, 1994). For other work on Silicon Valley, see Robert Kargon, Stuart W. Leslie, and Erica Schoenberger, 'Far Beyond Big Science: Science Regions and the Organization of Research and Development,' in Peter Galison and Bruce Hevly (eds.), *Big Science: The Growth of Large-Scale Research*, (Stanford, 1992); Stuart W. Leslie and Robert H. Kargon, 'Selling Silicon Valley: Frederick Terman's Model for Regional Advantage,' *Business History Review* 70 (Winter 1996), 435–72; Robert Kargon and Stuart Leslie, 'Imagined Geographies: Princeton, Stanford, and the Boundaries of Useful Knowledge in Postwar America,' *Minerva* 32 (Summer 1994), 121–43; Everett M. Rogers and Judith K. Larsen, *Silicon Valley Fever: Growth of High-Technology Culture* (New York, 1984); Michael S. Malone, *The Big Score: The Billion Dollar Story of Silicon Valley* (Garden City, 1985).
6. For a more detailed discussion of both the MOS transistor and Fairchild's work on it in the 1960s, see my dissertation, 'New Technology, New People, New Organizations: The Rise of the MOS Transistor, 1945–1975' (PhD dissertation, Princeton University, 1998).
7. Among the most prominent members of the R&D MOS group who left were Andy Grove, chief executive officer at Intel from 1987 to 1998, and Les Vadasz, as of April 1999 a senior vice-president and member of the board of directors. Some of the most research oriented of the Fairchild MOS group were either not invited to join Intel, or did not want to leave the R&D environment. Fairchild R&D had another branch of its MOS effort, which was developing a computer-aided-design system. Moore had lost faith in this work and did not continue it at Intel. Consequently no one from this group moved to Intel. In 1980, the key figures from this program started LSI Logic, which used computer-aided-design to build application specific integrated circuits. Rob Walker with Nancy Tersini, *Silicon Destiny: The Story of Application Specific Integrated Circuits and LSI Logic Corporation* (Milpitas: CMC Publications, 1992).
8. Gordon Moore, interview with author, 15 May 1996; Lee Boysel, interview with author, 23 January 1996. I have not been able to get an organization chart that shows the exact relation between the two groups.
9. A shift register is a memory element that stores data serially. For more on Wanlass and his role in the transfer of MOS technology, see Chapters 1, 4, and 5 of my dissertation.
10. Lee Boysel, interview with author, 23 January 1996; Lee Boysel, interview with author, 28 February 1997.
11. Gordon Moore recalled Boysel's secretiveness, remembering in particular being annoyed at one meeting where Boysel presented the completed design for a part that no one was even aware he was working on. Moore also acknowledged that Boysel had a better sense of how MOS technology could be used than anyone in R&D did. Gordon Moore, interview with Ross Bassett, 15 May 1996; Gordon Moore, interview with Ross Bassett and Christophe Lecuyer, 18 February 1997.
12. Lee Boysel, untitled document (13 September 1967) (in author's possession). The block diagram is reproduced on page 248 of my dissertation.
13. Ibid. For examples of work on MOS calculators, see *Electronic News* (21 September 1964), p. 8; Lewis H. Young, 'Uncalculated Risks Keep Calculator on the Shelf,' *Electronics*, 6 (March 1967), 231–34.
14. Those involved in Fairchild's initial development of integrated circuits knew the resistance a semiconductor manufacturer could meet when it tried to encroach on territory occupied by computer

designers. Gordon Moore related what happened when Fairchild went to one customer with its integrated circuit flip-flop, the basic storage element in a computer. According to Moore, the customer's reaction was:

> This is ridiculous. We need 16 different kinds of flip-flops. We have 16 engineers, each one of them a specialist in these flip-flops. There is no way we can use that single design for anything. It's a crummy flip-flop in the first place and it's not specialized for the things we need.

The customer's arguments ended when Fairchild offered to sell their flip-flop for much less than the costs of the components that made up a flip-flop. Gordon Moore, interview by Allen Chen, 6 January 1993, Intel Museum, Santa Clara, California.

15. John Hulme, Boysel's manager at Fairchild, had such a hard time communicating with Boysel that he considered him to be virtually in another company. Hulme also noted that some at Fairchild were suspicious of Boysel's motives. John Hulme, interview with author, 16 November 1996. Lee Boysel, interview with author, 23 January 1996; Lee L. Boysel, 'Memory on a Chip: A Step Toward Large-Scale Integration,' *Electronics* 6 (February 1967), 93–97; Lee L. Boysel, 'Adder on a Chip: LSI Helps Reduce Cost of Small Machine,' *Electronics* 18 (March 1968), 119–21; Lee L. Boysel and Joseph P. Murphy, 'Multiphase Clocking Achieves 100-Nsec MOS Memory,' *EDN* 10 (June 1968), 50–53; Fairchild Semiconductor, *MOS/LSI* (Mountain View: Fairchild Semiconductor, 1968).
16. Lee Boysel, interview with author, 28 February 1997; Lee Boysel, interview with author, 23 January 1996; John Hulme, interview with author, 16 November 1996. One reason Boysel was not sued by Fairchild, was the turmoil it was in at the time. In July 1968, Fairchild hired Lester Hogan from Motorola as president, who then brought a management team with him from Motorola. Motorola then sued Fairchild. Many Fairchild employees became disgruntled with the new management and left. In the resulting disarray Boysel's work may have been forgotten or not fully appreciated by the managers from Motorola.
17. Blyth & Co. 'Confidential Report on Four-Phase Systems, Inc.' (30 April 1971); Lee Boysel, interview with author 28 February 1997. Corning had an agreement with Boysel that if his company failed, he and his engineers would join Signetics.
18. Lee Boysel, interview with author, 28 February 1997. Four-Phase's early work is described in Marge Scandling, 'A Way to Cut Computer Costs?' *Palo Alto Times* 28 (April 1969), 9.
19. The relation between Four-Phase and Cartesian, which would now be called a silicon foundry, is an early example of the blurring of boundaries between firms that AnnaLee Saxenian has noted as a distinctive characteristic of Silicon Valley. Saxenian, *Regional Advantage*, pp. 29–57. Cartesian, Inc. 'Prospectus' (nd), in author's possession. While they were both at Fairchild, Cartesian's founder quoted Boysel prices per processed wafer which provided the basis for Four-Phase's early plans. These prices were hand written on a sheet of yellow paper. [Bob Cole] 'Price Per Wafer per design per order' (nd), in author's possession.
20. '"Intelligent" IV-70 Outguns IBM's 2260s and 3270s,' *Computerworld* 23 (June 1971), 26; Four-Phase Systems, Inc. 'Preliminary Prospectus' (30 May 1973).
21. Lee L. Boysel and Joseph P. Murphy, 'Four-Phase LSI Logic Offers New Approach to Computer Designer,' *Computer Design* (April 1970), 141–146. In fairness, when Intel introduced what is now known as a microprocessor, it did not have a stable name. The first thing Intel made that it called a microprocessor was a circuit board with a number of integrated circuits—this was a microprogrammed processor. Intel's early advertising called the 4004 both a computer on a chip and a microcomputer, but not a microprocessor. Because Intel was selling the 4004, it had to have a clear descriptive name for it, in a way Four-Phase never had to with the AL1. M. E. Hoff, Jr. and Stanley Mazor, 'Standard LSI for a Microprogrammed Processor,' *1970 NEREM Conference Record*, 92.
22. This paper will not cover the details of Intel's early microprocessor work. This has been addressed by participants, journalists, and historians in numerous papers and books. Federico Faggin, Marcian E. Hoff, Jr., Stanley Mazor, and Masatoshi Shima, 'The History of the 4004,' *IEEE Micro* (December 1996), 10–20; Tekla S. Perry, 'Marcian E. Hoff,' *IEEE Spectrum* (February 1994), 46–49; Robert N. Noyce and Marcian E. Hoff, Jr., 'A History of Microprocessor Development at Intel,' *IEEE Micro* (February 1981), 8–21; Federico Faggin, 'The Birth of the Microprocessor,' *Byte* (March 1992), 145–50; Stanley Mazor, 'The History of the Microcomputer: Invention and Evolution,' *Proceedings of the IEEE* 83 (December 1995), 1601–8; Malone, *The Micoprocessor: A Biography* pp. 3–20; William Aspray, 'The Intel 4004 Microprocessor: What Constituted Invention?' *IEEE Annals of the History of Computing* 19 (1997), 4–15.

23. Elizabeth de Atley, 'LSI Poses Dilemma for Systems Designers,' *Electronic Design* (1 February 1970), 44–52.
24. Four-Phase's advertising materials either did not mention anything about the AL1 or discussed it in general terms, such as putting 'the whole 75,000 component CPU on one card.' Four-Phase Systems, 'System IV/70: The Data Base Access System of the 70's' (8 January 1970) (in author's possession); Four-Phase Systems, 'Announcing System IV/70: The Data-Base Access System of the 70's' (no date), in author's possession.
25. Lee Boysel, Wallace Chan, and Jack Faith, 'Random-access MOS memory packs more bits to the chip,' *Electronics* (16 February 1970), 109–115. For the general environment regarding semiconductor memories, see pages 361–68 of my dissertation. I have not done the research to enable me to say that Four-Phase produced the first computer with an all semiconductor memory, but contemporaries generally accepted Four-Phase's claims in that regard. IBM ran an advertising campaign claiming its 370/Model 145 was the first computer with semiconductor memories. At an industry trade show, Four-Phase displayed modified versions of the IBM advertisements, claiming priority in semiconductor memory computers. *Electronic News* (12 October 1970), 57; Lee Boysel, interview with author, 23 January 1996.
26. Lee Boysel, interview with author, 28 February 1997; Lee Boysel, interview with author, 23 January 1996.
27. Hal Feeney, interview with author, 23 April 1997. Four-Phase's position in relation to Intel in semiconductor packaging, was similar to that of IBM compared with the semiconductor industry. IBM was willing to put much more money and effort into packaging than those in the semiconductor industry were. For a statement of the emphasis IBM as a computer company put on packaging, see Steven W. Usselman, 'IBM and its Imitators: Organizational Capabilities and the Emergence of the International Computer Industry,' *Business and Economic History* 22 (Winter 1993), 1–35.
28. Federico Faggin, 'The Birth of the Microprocessor,' *Byte* 17 (March 1992), 150.
29. Joel Karp and Elizabeth DeAtley, 'Use Four-phase MOS IC Logic,' *Electronic Design* 7 (April 1, 1967), 62–66. In Boysel's testing of the 1993 'microprocessor' version of the AL1 running a CTC application (the original user of the Intel 8008), the AL1 performed from 20 to 50 times faster than the 8008. Lee Boysel, 'AL1 Microprocessor Demonstration Model' (27 February 1993) (in author's possession). Although this paper stresses the differences between Intel and Four-Phase, there were connections. By going into standard LSI parts, (RAM, ROM, and later microprocessors) Intel was designing parts much like those Boysel had built at Fairchild. Robert Noyce was on the board of directors of Four-Phase and would have known the details of Four-Phase's work (and presumably taken it back to Intel). Blyth & Co., 'Confidential Report on Four-Phase Systems, Inc.' (20 April 1971) (in author's possession).
30. Eric von Hippel, *The Sources of Innovation* (New York, 1988).
31. D. L. Critchlow, R. H Dennard, S. E. Schuster, and E. Y. Rocher, 'Design of Insulated-Gate Field-Effect Transistors and Large Scale Integrated Circuit Chips for Memory and Logic Applications,' (4 October 1968) (in author's possession), 218. On Mead's reaction to the microprocessor, see Joyce Gemperlein and Pete Carey, 'If Hyatt Didn't Invent the Microprocessor, Who Did?' *San Jose Mercury News West* (2 December 1990), 25. For Mead's ideas at that time, see Carver Mead, 'Computers That Put the Power Where it Belongs,' *Engineering and Science* (February 1972), 4–9. For the IBM response, Dale Critchlow, note to author (19 July 1995).
32. In 1967 two researchers at IBM estimated that with LSI the industry might require as many as 100,000 unique part numbers. M. G. Fubini and M. G. Smith, 'Limitations in Solid-State Technology,' *IEEE Spectrum* (May 1967), 55–59.
33. Rob Walker with Nancy Tersini, *Silicon Destiny: The Story of Application Specific Integrated Circuits and LSI Logic Corporation* (Milpitas: CMC Publications, 1992); Gordon Moore, interview with author, 15 May 1996.
34. Intel's own brief experience with custom work shows its problems. It designed the 4004 in response to a request from Busicom, but Busicom never became a major customer, agreeing to give Intel greater rights to market the part in exchange for being relieved of obligations in the contract. Intel's 8008 was designed in response to a request from CTC which backed out of the project before it was completed.
35. Robert N. Noyce, 'Integrated Silicon Technology in the '70's,' *IEEE 1970 International Convention Digest*, 171; Robert N. Noyce, 'A Look at Future Costs of Large Integrated Arrays,' *FJCC* (1966), 111–114. Another former Fairchild manager expressed similar views. In 1968 Charles Sporck, the former general manager at Fairchild, who was by then the head of National Semiconductor said,

'Custom work kills a company when it's trying to grow. Henry Ford's Model T approach is just as valid now as it was then.' Quoted in Don C. Hoefler, 'Nat'l Semicon's Sporck Sees Firm Gaining Ground,' *Electronic News* (16 December 1968), 49.

36. The apparent irony is that Intel's microprocessors came from custom work for Busicom and CTC. But Intel had the strategy of using custom work to create standard parts. In 1970, Robert Noyce stated:

    Intel is actively soliciting custom business. We're doing this because we want to learn by working very closely with customers what they need to do their job. Hopefully by working with several customers in the same area, we can find the commonality that everybody seems to need, and then we can build that as a standard part. And once it exists as a standard part, the cheapest way for a guy to go will be to use it, because he will have all the advantages of a production line flow that is already established. quoted in Elizabeth de Atley, 'Can you build systems with off-the-shelf LSI?' *Electronic Design* 5 (March 1, 1970), 50.

37. Malone, *The Microprocessor: A Biography*, 3–20; William Aspray, 'The Intel 4004 Microprocessor: What Constituted Invention?' *IEEE Annals of the History of Computing* 19 (1997), 4–15. In interviews discussing the microprocessor, Ted Hoff has sometimes presented himself as the lone inventive genius, who had to struggle mightily to convince others of the worth of his invention, particularly those in marketing. But Gordon Moore claims that Hoff was under a misapprehension about who he needed to convince and claims to have immediately recognized the significance of Hoff's idea. Ted Hoff, interview with Rob Walker, Stanford University Library, Stanford University; Gordon Moore, interview with Rob Walker, Stanford University Library, Stanford University. Gordon Moore, interview with Ross Bassett and Christophe Lecuyer, 18 February 1997.
38. *Electronic News* (7 August 1972), 24.
39. Hal Feeney, interview with author.
40. Gordon Moore, interview with Stein, 17 October 1983, Intel Museum, Santa Clara, CA.
41. Paul Ceruzzi, 'From Scientific Instrument to Everyday Appliance: The Emergence of Personal Computers, 1970–1977,' *History and Technology* 13 (1996), 1–31.
42. Lee Boysel, 'The Way it Really Was,' 24 May 1974 (in author's possession). For a discussion of the effect of computer leasing on the computer industry see Katherine Davis Fishman, *The Computer Establishment* (New York, 1981), pp. 15–18; Franklin M. Fisher, John J. McGowan, and Joen E. Greenwood, *Folded, Spindled, and Mutilated: Economic Analysis and U.S. vs. IBM* (Cambridge, MA, 1983), pp. 191–96. Since the 1980s, the computer and semiconductor industries have switched, with the semiconductor industry so capital intensive as almost to preclude new entrants, while college students have started computer companies.
43. On the cover of its 1976 *Annual Report* Four-Phase had a Social Security operations center using its systems and asserted that the agency had acquired 1300 Four-Phase terminals. Four-Phase Systems, Inc., *Annual Report 1976*, pp. 1–2; Four-Phase Systems, Inc., *Annual Report, 1980*; *Wall Street Journal* (11 December 1981), p. 56; (4 January 1982), 33; (3 March 1982), p. 44.
44. Although those responsible for managing IBM in the 1980s and early 1990s have been pilloried for their performance, the case of Four-Phase may suggest why it took IBM so long to accommodate itself to the microprocessor. The economics were quite different for it as a vertically integrated producer of semiconductors that sold computers than they were for Intel. Although IBM has had microprocessor projects underway since roughly the time of Intel's initial work, the microprocessor has only relatively recently had a major role at IBM. (Now IBM sells semiconductors on the open market and has economic considerations closer to those of a semiconductor company.) For some of IBM's problems in this regard, see Charles H. Ferguson and Charles R. Morris, *Computer Wars: The Fall of IBM and the Future of Global Technology* (New York, 1994), pp. 30–97.
45. Gene Bylinksy, 'Here Comes the Second Computer Revolution,' *Fortune* (November, 1975) 184. One of the ironies associated with Intel is that although its staff consisted mainly of experts in semiconductor processing with very little computer expertise, it was able to transform itself into a computer company. But even today those roots are clear. All of Intel's chief executive officers as well as many senior managers have had backgrounds in semiconductor processing rather than computers.

Jon Eklund, Bernard Finn

Exhibition Critique: *Background for the* Information Age

The exhibit *Information Age* in the National Museum of American History had a gestation period of approximately five years—two of preliminary skirmishing, three of serious conceptual and development work. It was a team effort, involving five curators and a substantial supporting staff. The two of us were involved throughout the entire process. David Allison joined as lead curator at the beginning of the serious period. Uta Merzbach contributed to the initial phases, Steve Lubar to the final ones. The exhibit opened in May of 1990.

There are many advantages to a group effort. In this case one that stands out as particularly strong was the series of debates we had over the conceptual structure. The exhibit is, in our opinion, much richer as a consequence. There are also disadvantages. For us, most important was the fact that some of the debates were left unsettled, with the inevitable result that the final exhibit displays a number of loose ends that never got tied together.

We were determined, however, that the exhibit should speak through its artifacts, and we measure our success or failure largely in terms of how well we achieved that goal. We are therefore pleased to have the opportunity on these pages to present four instances (two each) where we think the objects served special purposes. Roger Bridgman then comments on their effectiveness.

Because the creation and sharing of information is central to society, it is reasonable to assume that a major change in information technology is likely to have a major social effect. In framing the exhibit we identified five general areas where technology is applied to an information signal: coding, processing, storing, communicating, decoding. We paid special attention to communications and processing, thus allowing us to use the contents of two of our strongest collections.

We organized the exhibit around the thesis that communications underwent a major 'revolution' when it became possible to convey information instantaneously (by electricity) in the 1840s (even though the practical impact of this technical ability would only gradually have an impact on society). Processing of information underwent a similar change a hundred years later with the invention of the electronic computer. For us the information age emerged in the 1970s when these two technologies, which shared an electronic base, effectively became one.

In addition to detailing the growth of these technologies, we tried to describe some of the major interactions with society: where social forces had had an impact on technology, and where technology had produced social consequences. For the latter we paid special attention to questions of who had access to the technology and who controlled it.

Such an analysis is hard enough to support in a historical monograph written for scholars, where evidence can be marshaled and exceptions noted. It is much more difficult in a popular exposition where words are at a premium, footnotes non-existent, and an attempt is made to have objects carry the bulk of the message. Nor did the team approach help very much. Nevertheless, it is our opinion that we succeeded more than we failed. We suggest that the four examples cited here are particularly good illustrations of what we had in mind.

Examples from Communications (Bernard Finn)

The telegraph instrument devised by Samuel F. B. Morse in 1835 and modified by him (with more windings on the magnet) in 1837 is mounted inside a case at the entrance to the exhibit. It is an American icon, especially impressive because of the crudeness of its construction and the openness of its design. As befits an icon, it is in an otherwise dimly lit area. But it is not alone. Next to it are its elements: for the receiver—an artist's canvas-stretcher similar to what Morse used as a supporting frame, an early electromagnet constructed by Joseph Henry, a contemporary wooden clock movement, a copper-wire paper-making screen from the early 19th century; for the transmitter—lead type and a printer's rule.

Beyond this assemblage are the more compact and professional-looking instruments constructed for Morse by Alfred Vail and used in the 1844 trial. And beyond those are five-needle and single-needle telegraphs of William Cooke and Charles Wheatstone (together with a Nobili galvanometer).

This group of objects, in a ten-meter-long case at the entrance to a major exhibition is supposed to convey four distinct messages. The Morse telegraph, by itself, is a symbol of the communications breakthrough made possible by electricity and the electromagnet. The device has an intriguing shape that should attract a visitor's attention, and the association with Morse's name (which is still recognized by most American school children) provides a strong focus for attention.

Second, together with the 'deconstructed' elements alongside, the Morse instrument tells about the circumstances of its invention. It was clearly made by someone with poor mechanical skills, with no money to hire assistance, using materials at hand. The value of luck is hinted at by the improved magnet, a critical element that was made known to Morse by a colleague only after the first instrument (in 1835) failed to transmit for more than a very short distance. And the type slugs and rule, which

Figure 1. Morse transmitter in exhibit. Photo by Laurie Minor.

Figure 2. Close-up of Morse.

are similar to what Morse would have seen every day as he went to his apartment above his brothers' printing establishment, suggest the origins of his 'portrule' transmitter.

Third, the Vail equipment underscores the importance of having a craftsman construct the equipment. Not only does it look better, it is 'obviously' more efficient.

Fourth, the Cooke-Wheatstone apparatus shows clearly that Morse was not alone in inventing an electromagnetic telegraph. It also indicates that with a different background his English rivals came up with a different mode of operation (though still dependent on Henry's multiple windings) borrowed from Nobili's measuring instrument.

Together, in addition to providing insights into Morse's invention, this unit—at the entrance to the exhibit—introduces visitors to the concept of the social influences on invention. (A companion unit, immediately following this one, is focussed on the Atlantic cable and emphasizes the social consequences of invention.)

About a dozen meters further on in the exhibit a display of objects from Alexander Graham Bell and Elisha Gray provide another opportunity to examine social forces on invention. In the years 1873–76 Bell and Gray pursued remarkably similar paths as they developed versions of a 'harmonic telegraph' (a form of multiplex telegraph where signals are sent at different frequencies) and then modified them so that they could convey speech. Bell, with a background of teaching speech, with no commitment to a career as an electrical inventor and with an intense desire to create something unique, saw in this fragile apparatus the makings of a commercial enterprise. Gray, an established electrical inventor, saw it rather as a curious device for demonstration of transmission of music, and he quickly returned to what he saw as more promising work on the telegraph. The museum has a wealth of material from both of these inventors, and the artefacts selected for display here show how closely their work proceeded in parallel.

The objects, of course, do not speak by themselves. Here, as in the Morse exhibit, supporting text and graphics are used to provoke visitors to look at the objects and to answer questions that might be raised by the objects.

Examples from Computers (Jon Eklund)

In the first half of the exhibit the visitor sees how electricity was employed in a succession of ways to create new modes of 'immediate' communications—telegraph, telephone, wireless, radio. Each in its own way was 'revolutionary,' in the sense that it emerged as an unexpected technology, looking for useful applications. In contrast, the processing techniques are seen as evolving in response to perceived needs—new filing technologies, non-technical methods of organizing the workplace, the Hollerith (punch-card) sorting machine, railway signaling systems.

Figure 3. Bell-Gray comparison. Photo by Laurie Minor.

Figure 4. Close-up of Bell-Gray comparison.

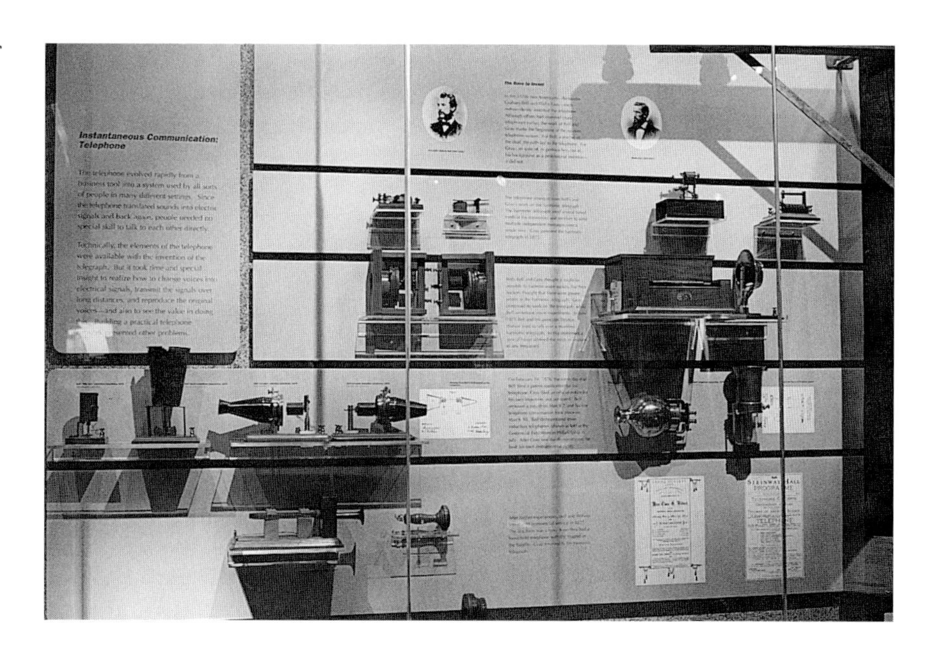

The breakthrough, the 'revolution' if you like, occurred when immediacy came to processing in the form of the electronic computer, and it is with such—ENIAC—that we start the second half of the exhibit.

Figure 5. ENIAC. Photo by Laurie Minor.

In contrast to the Morse instrument, ENIAC is large—taller than any of our visitors, and even the fifteen (out of an original forty) cabinets displayed are enough to fill a good-sized room. The machine is thus able, physically, to anchor this transition point in the exhibit. But it has several additional characteristics that should be meaningful to most observers. It is, literally, a 'black box' (or, more precisely, a series of black boxes); but at the same time it is open enough to display some aspects of its complex interior—thus representing all of the discreet-component electronic black boxes to follow, many of which will not be accessible. Some of these interiors include well over 500 vacuum tubes, an indication to a substantial number of visitors of the amount of power required and of the machine's vulnerability to tube failure.

Even the exterior, though black, is not without interest. The multiple dials continue the complexity argument, and their arrangement in rows and columns suggest the computational nature of the processes taking place inside. The plug-in connecting wires are an indication that even though the processing activity is virtually instantaneous, the programming activity is not. These messages are reinforced by accompanying text, but also by a video monitor showing portions of an interview with co-inventor Presper Eckert demonstration some of the features of the machine; there are also contemporary scenes of it in operation.

Figure 6. Bank of America Check Entry and Recorder. Photo by Laurie Minor.

Our fourth and final example is a 'check entry and recorder' from the Bank of America. It is an institutional-gray box with the electronics hidden. An operator reads the dollar amount written on the check, enters it using the keyboard (visible) so that it can be printed in machine-readable code on the check. This number, together with coded information already printed on the check, can then be transmitted electronically to other computers in the bank's system. The machine therefore represents the final transition point in the exhibit, where processing and communications functions become inextricably entwined, thus announcing the arrival of the 'information age.' The notion of transition is reinforced by a dozen cubby holes attached to the machine—vestiges of its earlier use (prior to modification) as a sorting machine.

Thus it is that we saw these artefacts when we chose them for the exhibit. Their ultimate effectiveness is obviously influenced by the design and the accompanying text. We have our own views of the degree to which these features facilitated or hindered the achievement of our goal. But instead of expressing them we look now for Roger Bridgman's opinions.

Roger Bridgman

Information Age—A Critique

I would like to start by making a few remarks about artefact-rich thematic exhibitions in general. My reason for doing so is that I will later voice some reservations about the ability of *Information Age* to 'speak through its artifacts,' and I would not like it to be thought that these reservations apply to *Information Age* alone. Indeed, looking critically at *Information Age* has pointed up the problems that afflict any exhibition trying to use artefacts as a communication medium. So in criticising *Information Age* I am questioning not so much an individual exhibition as the common beliefs from which it emerged.

The idea that you can use artefacts to tell a story enjoys widespread currency. But there are at least two obstacles in the way of those who, equipped with a storehouse of artefacts and impassioned by the knowledge of their history, would use the artefacts to communicate the passion. The first of these obstacles is the invisibility of unfamiliar artefacts, or at least the unfamiliar parts of unfamiliar artefacts: people find it difficult to see what they have never seen before. The second is that, even if an artefact is clearly seen, to use it in a story you have to use it as a sign, 'something which stands to somebody for something in some respect or capacity,'[1] and the meaning of any sign is socially determined. Neither of these obstacles exists for artefacts in common circulation; but these are the very artefacts that museums, as repositories of the exotic or the forgotten, tend to exclude.

If we cannot see what we have not already seen, how do we ever learn to see anything? As long ago as 1878 the German polymath Hermann von Helmholtz, who anticipated much of modern vision theory, drew attention to the role of knowledge in vision and offered an explanation of the way in which the seemingly impossible might be accomplished:

> If the similar traces, which are often left behind in our memories by repeated perceptions, increase, then it is precisely the law-like that repeats itself most regularly in a similar manner, while fortuitous change is eliminated. By this means there develops in the loving and attentive observer an intuitive image of the typical behaviour of the artefacts which interest him, of which he subsequently knows just as little as to how it came about as a child knows by which examples he has learnt the meanings of words.[2]

In the same lecture, Helmholtz went on to point out the necessity of this prior learning for vision, explaining that 'In present perception, the newly appearing sense impression forms the minor premise to which is applied the rule imprinted through earlier observations.' More recent theorists such as Richard Gregory likewise support the view that vision requires a knowledge base, however acquired.[3]

Everyday experience confirms this. Once you have looked at a few impressionist paintings, you see shadows on a cloudless day as blue. Before this experience, the same shadows seemed black. The battles surrounding those 19th-century painters who first saw blue shadows suggest that this new seeing was not straightforward. Indeed, anyone with formal art education will recognise the hours of painful effort that had to pass before they began to see artefacts with the completeness required for their convincing representation.

Most museum visitors arrive equipped only with the knowledge they need for seeing the familiar things of their world. Confronted with an artefact from outside it, say a siphon recorder,[4] they will see it only in their own terms. They will see the part that the curator (the epitome of Helmholtz's 'loving and attentive observer') sees as the siphon, for example, simply as a curiously shaped glass tube; they may not see its essential supporting thread at all. Until the visitor is more experienced, or a mentor intervenes, their perception remains at the cargo-cult level.

If there are problems with the purely visual, how much greater must be the visitor's difficulties in grasping the significance, not just the appearance, of unfamiliar artefacts. The need to read an artefact as a sign, when the language from which that sign is taken is unknown, is the second great obstacle in the path of telling stories with artefacts. Edwina Taborsky has argued against the 'observational paradigm' in which the meaning of the artefact is taken to exist in the artefact itself, not in the mind of the observer. This paradigm assumes that the cultural as well as the material content of an artefact is somehow 'put into' it by its creator and can be completely understood by anyone, even if they are not a member of the society in which it was created. When, as is usual, the process fails, this is attributed to 'misconceptions, and prejudices' in the observer's mind.[5] Umberto Eco dismisses as the 'referential fallacy' an even stronger, but frequently encountered, form of this paradigm, which regards an artefact as simply its material self, visible in a completely transparent way.[6]

Taborsky argues instead for what she calls the 'discursive paradigm', in which meaning arises from the interaction of artefact and observer at a particular place and time. In this paradigm, '[t]he artefact cannot exist as a sign or meaningful unit on its own.'[7] If you accept this—and I find it persuasive—it follows that the visitor's knowledge, as well as the artefact itself, is a key component in determining how the artefact functions as a sign. Though there is no single right meaning, something akin to the process of learning to see has to happen at the symbolic as well as the visual level if the visitor's encounter with an artefact is to generate the meaning the curator's story demands. If this learning has not happened, if the curator's and the visitor's worlds have not been aligned, the artefact–visitor interaction will generate the 'wrong' meaning and the story will falter.

Think of the siphon recorder again. The curator wants it to play a dramatic role in a story of human communication. But the visitor lacks the curator's knowledge of the vast network to which it was connected, the working lives of its operators, the way its siphon tube wagged magically from side to side as it traced out life-or-death messages. To the visitor, the instrument has meaning perhaps as a decorative antique, something that might make an aesthetic contribution to the home. The story is disrupted because interaction with the same sign creates different meanings for storyteller and listener.

The curator's traditional task is to repair this situation. By interweaving artefacts, text, perhaps even interactive exhibits, the curator first helps visitors to see artefacts more clearly, then imbues those artefacts with the symbolic values that come from their function and history. Then, and only then, are visitors able to make meanings that form a coherent story. In this way the museum fulfils its role of expanding visitors' mental world to include other cultures, other times, enhancing their understanding of their own culture and their own time. But look at the order of events: first the teaching, then the practice, then the moment when the exhibition can at last speak through its artefacts. By the time even an attentive visitor is ready for a story told by artefacts, the opportunity to tell that story has often passed: the visitor is heading for the exit.

All this implies that most visitors to artefact-rich exhibitions are in fact reading a story carried by text or other familiar media at the same time as they are learning to see artefacts and gathering new meanings for them. The artefacts, far from carrying the story, ride on its back. There is nothing wrong with this relationship, but pretending that it is the other way round can lead us to develop exhibitions that neither tell a clear story nor expand visitors' perceptions of artefacts. There is little doubt, for instance, that the best way of learning to see artefacts at the primitive level is to spend time looking at a lot of them gathered together in isolation.[8] In contrast, the best way to expand your interaction with them as signs is to explore them in a context that links them with other, more familiar signs like people and words. Museums started to shift from an artefact-intensive to an interpretive style in the early sixties, as the discursive paradigm began to gain ground from the observational. This approach, however, if taken too far, can impoverish the visitor's detailed perception to the point where almost any artefact could be substituted for any other without apparent incongruity. And neither style can solve the essential problem: if you do not already know the story behind an artefact, the artefact itself is powerless to tell that story.

My preamble done, I shall use some of its ideas in looking at the way five particular artefacts seem to function in *Information Age*. Morse's telegraph, Bell's and Gray's telephones, ENIAC, and the Bank of America ERMA cheque entry and recording machines. Do these artefacts succeed in speaking for themselves?

The problem is perhaps presented most clearly in the Morse telegraph exhibit. From its position at the entrance to the exhibition, this is clearly intended to have iconic impact, making the most of its associations with a name every schoolboy knows and with the dawn of electrical communication. But there is also a very proper desire to explore the humble origins of this groundbreaking device; to this end, it is deconstructed, ostensibly to reveal it as an assemblage of homely odds and ends. To see whether this game plan can succeed, consider what the visitor needs to know before the artefacts can start to do their work. First, what a telegraph is. Second, that Morse invented the first viable telegraph system. Third, that he was a painter and his brother a printer. And finally, what sort of paraphernalia you might expect to find lying around an artist's studio and a printer's shop.

Artefacts, or at least working models, might be able to say what a telegraph is, though you would still need the word 'telegraph' as a link between artefact and further discourse. But about Morse, his achievement, his profession, his family connections and the identity of the artefacts you see, the artefacts themselves are silent. Only text can tell this part of the story. And though a canvas stretcher, the wire of a paper mould and a composing stick might seem to be artefacts that can speak for themselves, eloquent in their ordinariness, this is probably a curatorial illusion. All of these minor artefacts, to the eye of the non-specialist visitor, are likely to be as obscure, hard to see and lacking in meaning as the major artefact that they are intended to illuminate. Most people don't paint in oils, use copper wire or set type any more than they communicate by telegraph. In contrast, the clock will be clearly recognisable and laden with meaning (at least for older visitors), showing that the deconstruction could have worked if it had happened to produce more meaningful fragments. One could perhaps say that at least the crudity of Morse's early equipment is evident; but what counts as crudity to innocent modern eyes? Today's child probably sees everything more than a century old as crude in the extreme. Text comes to the rescue again, of course, and everything is made clear in the end; but it is the end, not the beginning. Most of the artefacts have not spoken—they have been spoken about.

Bell's and Gray's telephones present further difficulties. As more sophisticated devices, they do not even lend themselves to deconstruction into components. They have to be seen whole. Once again, the difficulty of seeing arises. Though these are telephones, they do not look anything like the artefact that the word 'telephone' commonly signifies. And though visitors will accept, once told, that they are indeed strange-looking telephones, their perception will at first be confined to simple shapes and textures. These will not, without further study, congeal into the whole, functional artefacts seen by the curator's eye. Much less will they declare themselves as the product of Bell's or Gray's workshop. The visitor simply has to be told: this one is Bell, this one Gray.

A thought experiment that I like to carry out at this point, to check whether text is doing its job of supporting and supplying meanings for artefacts, is to swap the artefacts around. Would anyone but an expert notice? Does the text tell ordinary visitors what to look for, and connect these diagnostic features to the stories of the artefacts' creation and use? If not, an opportunity to enlarge visitors' repertoire of meanings has been missed, and it will remain difficult to tell stories with these artefacts. I did feel in this section that the artefacts are used as illustrations to a textual story, neither telling the story themselves nor taking full advantage of the text to grow in meaning. One message that visitors might be expected to get from the very similarity of these particular artefacts, for example, is that Bell's and Gray's paths of development were highly convergent. But unfortunately this needs pointing out in words: to many visitors, all this material looks very much the same anyway. In a world without firm points of reference, how similar is similar?

With ENIAC we reach another level of difficulty. The artefact is so large that you see not ENIAC but a part that stands for the whole. The exhibition repairs this common and unavoidable defect with a splendid collection of video material showing the complete machine. But in consequence the portion of ENIAC that is on display is reduced to the status of a relic. Like a splinter of the true cross, it cannot tell its own story; its job is simply to recall and reinforce belief in that story for people who already know it. After they have read the text and watched the videos, visitors too will know the story of ENIAC, but it is those supporting media, not the artefact, that will have told them it.

At least ENIAC, though, as a more recent artefact than Morse's telegraph or Bell's telephone, has a few features that visitors may recognise. It has the modular layout, and even the black finish, now familiar in hi-fi equipment. Its plugs and sockets look very like the mess you see at the back of your PC. And its function table is, on a gigantic scale, quite like the PROM that has long since replaced it (though only a few visitors would recognise that). These are points at which the artefact could, simply by being placed beside such modern artefacts, speak to at least some visitors. The story it would tell, though, would not be the story in the text, which is perhaps why the links were not made at this point, even though the exhibition talks about these subjects elsewhere. The lesson here, perhaps, is that artefacts do always tell a story of some kind; but, until text or other media intervene, that story is woven out of the stock of perceptions and meanings that visitors bring with them. Curators may feel that the replacement of this naive story by received history is their primary task. Maybe it is, but it will always be difficult to do it with artefacts as the primary medium.

And what about artefacts that, rather than belonging to a single point in time, span a timeline of their own? Like old houses that have been altered over the centuries, they are records, not just representatives, of change. They offer a correspondingly more complex set of signs, and untangling their meanings can present a challenge even for the experienced observer. The two Bank of America ERMA machines—the 'proof machine' for manual entry of dollar amounts and the 'dollar-amount encoder' for printing the sum on the cheque—fall into this category. While their appearance, as relatively modern artefacts, is not on the level of perceptual difficulty presented by Bell's telephones, their function as signs is not equally simple.

An attentive visitor might notice the set of bins forming part of the proof machine and make the link to Hollerith's card-sorting equipment. This would indeed be an artefact talking. But what exactly would it say? The obvious message, that ERMA was a punched-card system, is false. The true message, that the machine evolved from one used on an earlier system that did employ punched cards, is not obvious. And the deeper message, that the machine could transmit its data to other computers in the system, heralding the Information Age of the exhibition's title, can, in the absence of the relevant artefacts, be conveyed only by text.

Yet *Information Age* is far from deserted. Its mix of text and artefacts, on whatever level it works for the individual, clearly succeeds in bringing in visitors for an experience that is in some way richer than that given by other media. Text and artefacts, in short, have been skillfully mixed to make a formula more powerful than either ingredient alone. Some participants in the resulting complex museum experience will be naive observers, acquiring early exposure to historic artefacts; a minority will be mature museum-goers. *Information Age* serves the inexperienced well, with text that is clear and lively; they will lean on this, using the artefacts as a kind of historical decoration to a verbal message, at the same time learning to see them better. The exhibition also provides much for the experienced, who can enjoy a profusion of important artefacts that it would be difficult to match elsewhere. For them, the text provides interesting commentary rather than essential information.

The artefacts that *Information Age*'s creators have selected for discussion in this paper are well chosen. They span a range of possibilities and problems in attempting an exhibition that can speak through its artefacts. Some, whose language is a forgotten dialect, are mute. Others speak, but tell unwanted stories. Yet others seem to lie. In all cases the curator's solution has to be the same: to let human language carry the burden, relegating the artefacts to the role of illustrations or relics. The text carries the artefacts, enriching them as perceptions and signs. Only a few visitors to *Information Age*, as to any other exhibition of its kind, will be able to reverse this relationship and use the artefacts to enrich the meanings of the text. This favoured minority will already have absorbed the visual language and cultural meanings of the relevant class of artefacts through exposure to taxonomic displays, text-mediated interpretative exhibits and other sources. Like everything else, museum visiting needs practice; and visiting museums where artefacts do the talking needs the most practice of all.

Notes

1. C. S. Peirce, *Collected Papers*, edited by P. Weiss, C. Hartshorne and A. Burks (Cambridge MA, 1931–58), Vol. 2, Sec. 228.
2. Hermann von Helmholtz, *Science and Culture: Popular and Philosophical Essays*, edited by David Cahan (Chicago, 1995), p. 355.
3. See J. Anderson, H. B. Barlow and R. L. Gregory (eds.), 'Knowledge-Based Vision in Man and Machines,' *Philosophical Transactions of the Royal Society of London: Biological Sciences*, Vol. 352, No. 1358 (29 August 1997).
4. A form of telegraph receiver in which ink, siphoning through a narrow tube actuated by an electromagnet, marks a moving strip of paper. Those who did not recognize the term may have experienced, in textual form, something of the non-seeing alienation of the naive observer.
5. Edwina Taborsky, 'The Discursive Artefact' in Susan Pearce (ed.), *New Research in Museum Studies* (London, 1990), p. 60.
6. Umberto Eco, *A Theory of Semiotics*, (Bloomington, 1976), pp. 58–62.
7. Taborsky, 'The Discursive Artefact,' p. 58.
8. The Heinz Nixdorf Museums Forum in Paderborn, Germany, has, in its artefact-rich but basically text-mediated exhibition, a splendid display of what looks like every pocket calculator ever made. Curator Ulf Hashagen has made little textual comment here, and none is needed. One sees; one learns to see.

Paul Ceruzzi

'The Mind's Eye' and the Computers of Seymour Cray

A few years ago I spent a summer at the Computer Museum, which at the time was in the process of opening its public space on Museum Wharf in Boston. One day the Director, Dr. Gwen Bell, asked me to go to a back room, remove a circuit board from a computer there, and bring it out in preparation for display. I was accompanied by a summer intern, who by that time had already gained a lot of hands-on knowledge of the museum and its (at the time) unusual collections. The computer was a 'Naval Tactical Data System' (NTDS), built by Sperry UNIVAC for the U.S. Navy. When we opened it up, the intern looked at me and said, 'It's pretty obvious that Seymour Cray designed this machine, isn't it.'

He was correct. Seymour Cray was involved in the initial work on the NTDS. The Naval Tactical Data System was a specialized computer built for the U.S. Navy. It was not well-known outside the small and highly-specialized military market. It was built by Sperry UNIVAC, a company that few people even today know that Seymour Cray ever worked for. Cray left Sperry UNIVAC by the time the final product was designed and built, and there was little in the published record that documented his involvement with this machine. His name appeared nowhere on the machine, nor in any technical manuals or other descriptions that we had of it at the time. So what did the intern know, and how did he know it?

This paper is an attempt to answer those questions.

Most curators involved with museums of technology are familiar with the notion of visual, as opposed to verbal, 'readings' of technology, as developed by Eugene Ferguson in his influential article on 'The Mind's Eye,' first published in 1977 and later expanded into a book.[1] Curators who collect and exhibit artifacts related to microelectronics and computing find that, perhaps, those ideas may not be as helpful as they are when dealing with earlier technologies, whose mechanisms are more easily seen and grasped by the observer. One obvious clue to this problem is the term often used when discussing electronics: black box. The implication seems to be that at some point one turns away from an attempt to understand what is going on inside.[2]

This is of course a much-condensed statement of the issue of black boxes, and I hope I have not done too much harm by summarizing it so briefly. I do believe it is not a distortion of the reality faced by many curators in museums of moden technology, who face the task of developing exhibitions on computing and electronics for a mass audience.

Figure 1. UNIVAC NTDS Computer, ca. 1962. The NTDS was built for installation on board navy ships, and so was housed in a rugged, water-tight box. UNIVAC also marketed a commercial version, which was electrically identical but packaged in a more conventional housing.

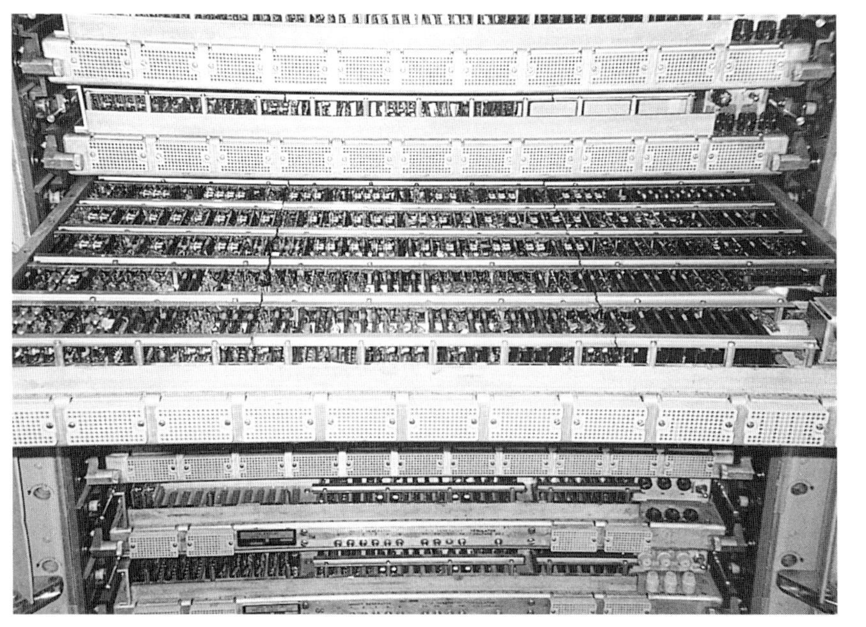

Figure 2. Close-up of the NTDS Circuits. The dense packing of the individual circuits was an indication that Seymour Cray was involved. The NTDS used discrete transistors, not integrated circuits. (Source for both photos: Dag Spicer, the Computer Museum History Center, Moffett Field, California.)

Figure 3. Intel 8080a chip, ca. 1975. The 8080 was used in the early personal computers and led to the 8086x series that still dominates personal computing. Like most integrated circuits, the chip itself is encapsulated in a black, rectangular housing, with electrical contacts protruding from two sides in parallel rows. This so-called 'Dual In-line Package' (DIP), invented around 1970, remains in common use in the electronics industry, in spite of enormous advances in chip density. (Photo: Smithsonian Institution, 92–7008–29)

Although I feel as passionately about these artifacts as those who study older, classical machines, I seldom experience the kind of pleasure Ferguson describes when analyzing the design, say, of a motorcycle engine. For the past 25 years—an eternity in computing history—digital computers have been made of small, black, rectangular chips, soldered onto rectangular printed circuit boards, which in turn are plugged into a bus, or backplane, that supplies power, communication signals, and so forth. The whole arrangement is housed in one or more rectangular boxes. Not much to read, it seems.

The intern's reading of the NTDS computer suggests an exception to this state of affairs.

Seymour Cray (1925–1996) was a legend in the computer industry for his design of supercomputers—machines that could calculate much faster than any others. Supercomputers were eagerly sought by government and defense agencies such as the U.S. National Security Agency, the Lawrence Livermore Labs, and NASA. Their export to other countries, even those friendly to the U.S., was restricted, although that policy was probably as much a matter of politics as it was a matter of technology. For this discussion, more relevant than the politics of its use or the work it did was that Cray computers designed after 1976 were circular or semicircular in shape—a rare exception to the rectangular boxes that characterize all other computers, large and small.[3]

As Donald MacKenzie and Boelie Elzen have pointed out, the success of the CRAY-1 was due to a complex mix of factors, technical, social, and political.[4] I have already mentioned the political factors that restricted the export of these machines; to that MacKenzie and Elzen add the cult of personality that surrounded Seymour, a result of his Midwest roots and demeanor, his distaste for bureaucracy, his refusal to talk to the trade or

Figure 4. Seymour Cray, standing next to a CRAY-1 computer, ca. 1976. The naugahyde-covered power supplies are visible in the lower part of the photo. Each segment of the CRAY-1's processor had its own dedicated power supply. (Photo: Cray Research, Inc. SI photo # 89–21494).

Figure 5. A typical installation, with the processor surrounded by a 'farm' of disk drives that make the room look a bit like a laundromat. (Photo: Cray Research, Inc.).

popular press, and finally, the unique shape of his designs. (Part of the cult is that people usually refer to him by his first name.)

In one of his rare public appearances, Cray described how he arrived at that shape as a result of a desire to minimize the time a signal would spend traveling from one end of the computer to another. The interior, containing connecting wiring, had however to be large enough for a small-framed person (typically a woman) to squeeze inside and make a repair if necessary. Power supplies had to be located close to each set of circuits, to minimize power losses. Thus he arrived, for what he described

Figure 6. Wiring a CRAY backplane. Cray computers were laboriously handwired, to give the fastest possible speeds for signals to travel from one part of the computer to another. A) Seymour Cray had a plant built near his home in the remote town of Chippewa Falls, Wisconsin to do this final assembly. He stated that it was easy to find, hire, and train a highly-capable workforce, mainly women, from the surrounding region to do the work (Credit; Cray Research, Inc. SI photo # 89–21495).

as purely rational reasons, at the three-quarters circular column, with a lower bench containing the power supplies. The padding on the bench was perhaps his only concession to whimsy. Although the machine has been called the 'world's most expensive love-seat,' the owners of a multi-million dollar machine would seldom allow people to sit there unless they had a good reason do so.

A closer look at the design suggests that perhaps it was not so rational after all. The circle was not closed; thus the time it would take for a signal to travel from one end to the other was only slightly less than it would have been had the circuits been laid out on a rectangular frame. Except for the CRAY-2, other machines produced by Cray Research, including the CRAY X-MP, Y-MP, and all subsequent computers, either were rectangular or had only a vestigial reminder of the semicircular shape. That may have been for marketing as well as for technical purposes. The computers that Seymour designed before founding Cray Research, including the NTDS mentioned above as well as the Control Data Corporation CDC-6600 (his first supercomputer) were all rectangular.

What really distinguished all of Cray's designs, and what enabled the intern to recognize his early work so readily, was that Cray succeeded in packing circuits to a density that no one else could approach. For his pioneering CDC-6600, Cray packed the resistors, transistors, diodes, and other components so densely they resembled stacks of firewood; the process was called cordwood packaging. To do that he had to face and solve a daunting problem. Electronic components give off heat, but in a dense package there is little room for the heat to escape. Cray came up with innovative liquid cooling techniques that, in may respects, were the key to his success with the CDC-6600 and the CRAY-1. Among computer engineers, Seymour is remembered as much for the genius of his packaging and cooling techniques as he is for his brilliant mastery of electronic circuit design. The CRAY-2 was completely immersed in a bath of liquid to keep it from overheating.

A person with some familiarity with packaging can thus recognize a Seymour Cray machine. One can do that even though his innovative logic, which he incorporated into the circuits and which is the stuff that goes on inside the chips and other 'black boxes,' remains invisible. It may be an exceptional case, but in at least this instance one can 'read' a computer, and museum curators can design public exhibits that exploit this.

That leads, finally, to a mystery that I encountered when I recently acquired a computer from the U.S. Air Force. The computer was a Control Data CDC-3800, and it was offered to the National Air and Space Museum by the Air Force's Onizuka Air Force Base, located in Sunnyvale, California. When I received this offer I accepted it almost immediately. That was not because I knew anything about the computer—I didn't—but because of the place where it was located. I knew

Figure 7. A CRAY-2, in front of its disk farm. The tubes in the front of the photo held the inert cooling liquid, which flowed through and around the processor circuits. (Credit: Cray Research, Inc.)

Figure 8. A CRAY-1, serial #14, on display at the National Air and Space Museum, in a gallery devoted to the use of computers in aerospace. (Photo: Smithsonian Institution, # 92–15054–6)

however that Onizuka Air Force Base, named after an astronaut killed in the *Challenger* accident, was a base in name only. In fact it is a large, rectangular, blue building, whose very existence was classified throughout the Cold War. It remained secret in spite of the fact that the building is right next to a busy freeway in the middle of Silicon Valley, and that investigative journalists had already described the building and its functions, giving it the popular name the 'Blue Cube.'[5] It was the place where U.S. intelligence satellites were controlled and operated. The very existence of these programs was kept classified throughout the Cold War years. Even today these activities remain among those classified as black: their budgets are buried within other appropriations, to make it difficult for anyone without the proper clearance to know what is going on.

The computer being offered to the Air & Space Museum had been involved in these activities.[6] Given the importance of satellite reconnaissance, and the scarcity of any artifacts related to it in the Smithsonian's collections, it seemed appropriate to accept this offer regardless of whatever technical innovations the CDC-3800 had or did not have. We knew

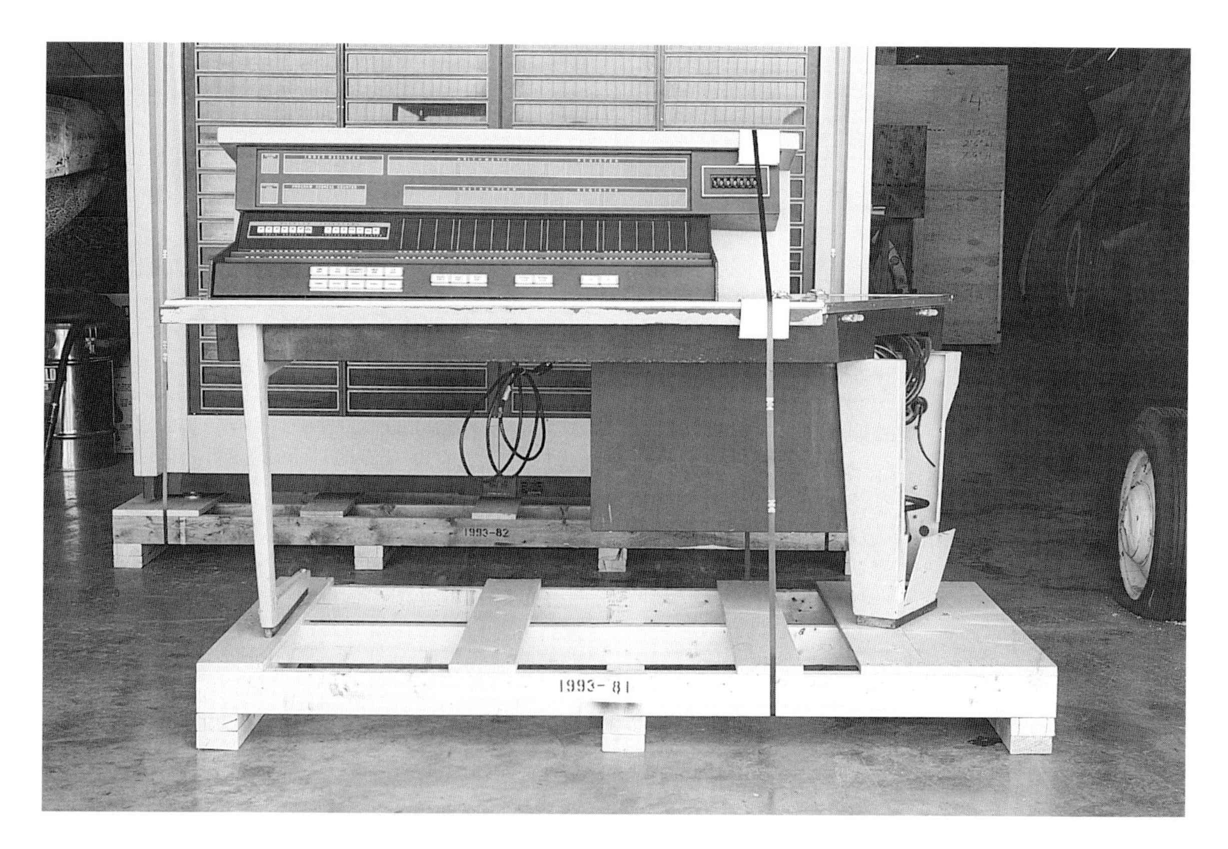

Figure 9. The Air Force's CDC-3800, now in storage at the National Air and Space Museum's Paul E. Garber Facility in Silver Hill, Maryland. Plans are to exhibit the computer, in the context of reconnaissance satellites and other military programs, at the museum's new facility at Dulles Airport. (Credit: NASM)

how difficult it was to exhibit computer hardware in our museum. We also knew that the public was familiar with these activities, through venues like Tom Clancy novels and Hollywood movies starring actors like Harrison Ford. These fictitious accounts are far removed from what really happens, but they do provide a way, however imperfect, to bring a visually-uninteresting object to life.

As might be expected, I received little in the way of documentation when I acquired the computer.[7] Nor was I able to find much about it in the published descriptions of Control Data products, although information is available about its civilian counterpart, the CDC 3600. An informal query posted on an Internet discussion group turned up several retired CDC employees who believe that Seymour Cray was involved with the design of the CDC-3800, but they are not certain. The one technical description of the CDC-3600 that was published at the time of its introduction, however, indicates that it had a different architecture from the Cray-designed CDC-6600, and that it was related to CDC computers that Seymour had nothing to do with.[8]

So I conclude with a question: did Seymour Cray design the CDC-3800? Cray was employed at CDC at the time, and he could have been involved in its design at some stage. I have done some preliminary research, but I have not had an opportunity to do more detailed archival research that might establish its pedigree. But to harken back to the Computer Museum intern's reading of the NTDS: can one tell simply by looking at it? The computer is housed in a series of rectangular boxes. The circuits are densely packed, with wire-wrapped connections among the circuit boards. The frame is massive—the nameplate alone weighs as much as the computer on my desk! Does that make it a Cray machine? Current plans are to exhibit the computer as part of the opening of the Air and Space Museum's Extension, at Dulles Airport, sometime around 2003. When that exhibit opens, I hope to develop this theme in the labels for the computer, and thereby bring the visitor into the topic. By then I expect to have looked at the archival record and thus know the answer based on written texts. Regardless of what I find, I intend to keep a reading of the text of the machine itself.

Notes

1. Eugene S. Ferguson, 'The Mind's Eye: Nonverbal Thought in Technology,' *Science*, 197 (26 August 1977): 827–836. Relevant to this paper is the fact that when Ferguson revised and expanded that paper into a book, *Engineering and the Mind's Eye*, (Cambridge, MA, 1992), he added a discussion about the introduction of computers, especially of the practice of computer-aided-design (CAD), expressing concern that insofar as it reduced emphasis on the non-verbal, CAD was causing the practice of engineering great harm.
2. As in, e.g. the well-known series of books by Nathan Rosenberg: *Inside the Black Box*, (1982), and *Exploring the Black Box*, (1994).
3. Donald MacKenzie, 'The Influence of the Los Alamos and Livermore National Laboratories on the Development of Supercomputing,' *Annals of the History of Computing*, 13/2 (1991): 179–201.
4. Donald MacKenzie, *Knowing Machines: Essays on Technical Change* (Cambridge, MA, 1996), Chapter 6, with Boelie Elzen.
5. William E. Burrows, *Deep Black: Space Espionage and National Security* (New York, 1986). Burrows says very little about what happens in the 'Big Blue Cube,' which he identifies as the Air Force's Satellite Control Facility. Since the end of the Cold War other more detailed accounts have appeared. Beginning around 1990 the operations at the facility were transferred to a 'Consolidated' Space Operations Center in Colorado, although Onizuka AFB remained active. Burrows claims that the reason for the transfer was that the facility is located on an active earthquake fault, and it is vulnerable to a terrorist who could simply drive up to it along U.S. Highway 101. I do not find either argument convincing.
6. James B. Schultz, 'Inside the Blue Cube: USAF Modernizes Satellite Tracking Network,' *Defense Electronics* (April 1983): 52–59.
7. I was allowed inside the facility, where I saw the computer installed in an ordinary mainframe computer room, just prior to its being declared surplus. I do not have a security clearance.
8. Charles T. Casale, 'Planning the 3600,' *AFIPS Conference Proceedings*, vol. 22, Fall Joint Computer Conference (1962), pp. 73–85. A talk given by C. Gordon Bell in November 1997, which he posted on the World Wide Web, stated that 'Cray worked on the circuitry for the 3600 en route to the 6600.' (http://www.research.microsoft.com/research/barc/Gbell/craytalk/index.htm)

Paul Forman

Researching Rabi's Relics: Using the Electron to Determine Nuclear Moments Before Magnetic Resonance, 1927–37.

Through the last two decades of his life I. I. Rabi had insisted that nothing remained of the apparatus employed in his molecular beam researches prior to the Second World War: 'It's all gone.'[1] When, however, following his death in January, 1988, Rabi's papers and effects were removed from his office at Columbia University, three pieces of 'hardware' came to light and were donated to the Smithsonian Institution's National Museum of American History (Figure 1).

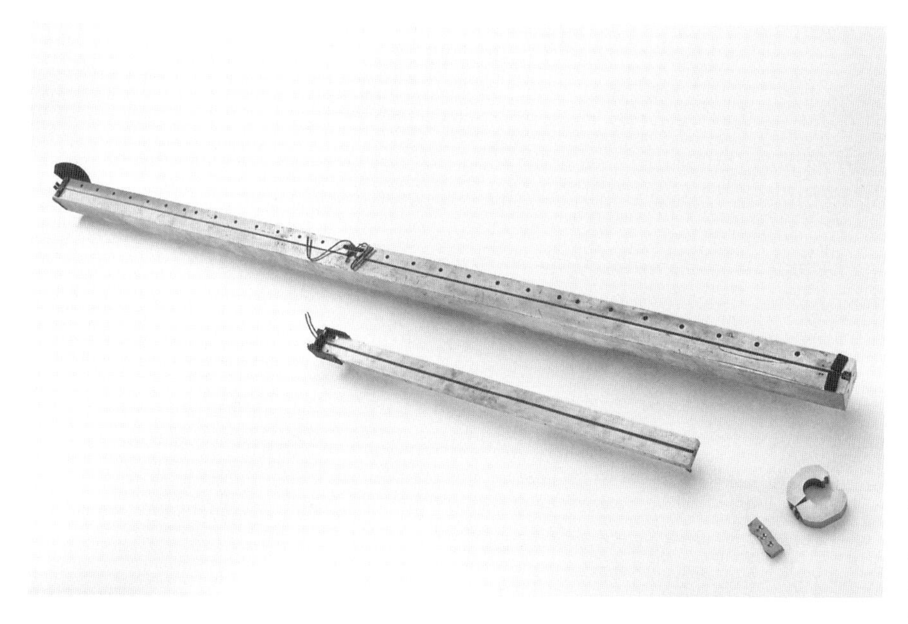

Figure 1. Three molecular beam deflecting magnets as received from the estate of I. I. Rabi early in 1989 (National Museum of American History accession no. 1996.0331). The two bars, of length two feet and five feet, precisely, are of non-magnetic aluminum alloy. Each carries, embedded in one side, a pair of thin copper tubes. The object at lower right is a C-shaped (annular), ferric magnet, formed of two opposed halves, here hinged together and partially opened. These three magnets will be denominated, respectively, the 'Millman' magnet, the 'Manley' magnet, and the 'indium' magnet. In use, the bars were rotated 180° from their orientations in this photograph (see Figure 3). The hinge on the 'indium' magnet is attached at the 'bottom' of the magnet, whose gap, in use, was vertical at top (see Figure 4).

All three of these objects are molecular beam deflecting magnets employed in measurements of angular momenta and magnetic moments of atomic nuclei, measurements carried out under Rabi's direction at Columbia over the five years 1933–37. The technique employed in those measurements, devised by Rabi and Gregory Breit in 1930/31,[2] was an ingenious modification of that first introduced by Otto Stern in 1921 (Figure 2).

In a lengthy paper being published in sequential parts in *Annals of Science*, I report an investigation—in which the assistance of Roger Sherman, Museum Specialist in the Electricity and Modern Physics

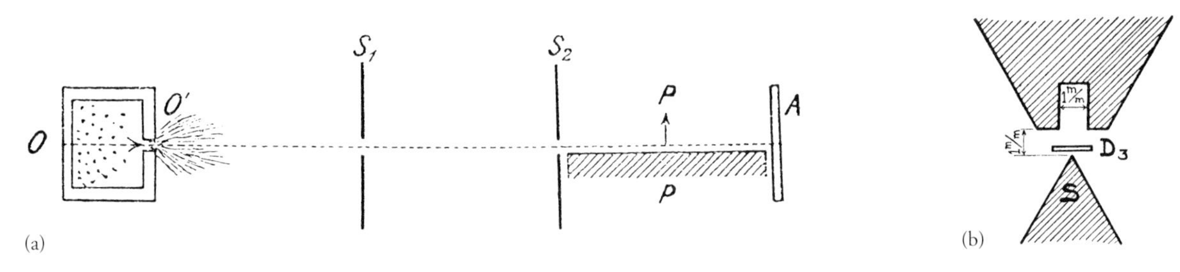

Figure 2. Schematic drawings of the Stern-Gerlach experiment. At left (Figure 2a) the substance to be investigated is vaporized in the oven, O. Evaporated atoms (or molecules) emerge into an evacuated chamber at O', and are formed into a ribbon-like beam by the slits, S_1, S_2, whose long dimensions are perpendicular to the paper. The beam so formed is then deflected by the action of an inhomogeneous magnetic field, between the pole pieces PP, upon the magnetic moments (electronic and/or nuclear) of the atoms. Finally, the atoms strike and adhere to the arrester plate, A. The separations between the traces left in the presence and in the absence of the inhomogeneous magnetic field are a measure of the strength of the magnetic moments borne by the atoms. At right (Figure 2b) is a transverse section through the pole pieces, showing the configuration Stern and Gerlach employed to maximize the magnetic field and its gradient—and hence to maximize the deflection of the atoms. Figures reproduced from: Walther Gerlach, 'Über die Richtungsquantelung im Magnetfeld II. Experimentelle Untersuchungen über das Verhalten normaler Atome unter magnetischer Kraftwirkung,' Annalen der Physik 76 (1925), Figure 1 on p. 164; Gerlach, 'Der experimentelle Nachweis der Richtungsquantelung im Magnetfeld und das magnetische Moment des Silberatoms,' Physica 2 (1922), Figure 2 on p. 123.

Collections, has been essential—aiming to describe and identify those three objects and to explicate them as historical artefacts within Rabi's developing program of research.[3]

Making use of the documentary resources available, especially Rabi's manuscripts, correspondence, and projection slides deposited at the Library of Congress, I provide in that longer paper a detailed, critical account of Rabi's route into molecular beam research, and of the first decade of his work with that technique. I point out Rabi's early commitment to the field of magnetism, highlight the extent of his indebtedness to Otto Stern's pioneering efforts toward molecular beam measurements of nuclear moments, and, especially, make evident the coherence of the research program for determination of nuclear moments by the 'Breit-Rabi method' that Rabi pursued from 1930 to 1937, i.e., prior to his implementation of the justly famous magnetic resonance method. Here in this brief paper I scant biographical and institutional circumstances and offer a very condensed account of that research program and the roles of these objects within it.

Born Israel Isaac in 1898 in a small town in Galicia—born to poor, little educated, deeply religious parents in this proverbially impoverished Polish province of the Austro-Hungarian Empire—Isidor Isaac Rabi spent his childhood and youth in immigrant neighborhoods of New York City. A voracious reader, exceptionally bright and broadly gifted intellectually, Rabi was long uncertain of his vocation, but found his way by age 30 into 'the purest kind of pure science.'[4] Such, for Rabi, was the technique of molecular beams, as he found it practiced by Stern at Hamburg University. From 1928 Rabi took this as his own field of research, applying it more and more exclusively to the measurement of the spins and magnetic moments of atomic nuclei. A decade of significant conceptual and instrumental advances, advances with which the artefacts here in question are closely associated, was then quite overshadowed by that further improvement in the technique—the magnetic resonance method—that Rabi and Co. began to employ late in 1937. And 'magnetic resonance' so increased the sensitivity and precision of the molecular beam method as largely to obliterate all earlier advances from scientific memory.[5] But for Rabi himself those earlier researches—and the magnets he had designed for them—remained an important part of his own sense of accomplishment and of his place in the history of science.[6]

Rabi was not wrong so to regard himself and these artefacts. In the late 1920s nuclear physics was becoming widely recognized as the 'new frontier' of physical research, and in the early 1930s American physicists, more than those of any other nation, began turning their research efforts in this direction. Over the six years 1932–1937, in which Rabi's pre-magnetic-resonance researches were appearing in the *Physical Review*, nuclear physics was springing from under 10% to over 30% of the papers published in that journal.[7] Thus Rabi, a bit of a Rutherford, was on the crest of a wave that he himself was helping to create. Moreover, what he contributed to the making of that nuclear physics wave was in some respects unique. Nuclear moments—i.e., a nucleus's 'spin' (angular momentum) and its magnetic moment—were among the very few parameters then regarded as necessary to define a nucleus as a quantum-physical system. While optical spectroscopists, through their analyses of hyperfine structure, provided most of the experimental data on nuclear moments, Rabi and only Rabi was providing confirmation and supplementation of that data by an independent technique, the Breit-Rabi method.[8] Those pre-magnetic resonance experiments were imaginatively conceived, skillfully performed, knowledgably analyzed—and technically demanding. Experimentalists so able and successful as Ernest Lawrence listened 'almost with reverence' to Rabi's account of them.[9]

Measuring nuclear magnetic moments was a desideratum that Rabi's principal mentors had earlier emphasized: Ralph Kronig, with whom Rabi studied and collaborated at Columbia University, January 1926 to June 1927; Wolfgang Pauli and Otto Stern, at Hamburg University, October 1927 to December 1928.[10] The question took on great importance through the acceptance early in 1926 of the concept of electron spin—the attribution to every electron of an intrinsic angular momentum of magnitude exactly half that possessed by an electron circulating in the lowest orbit of a Bohr atom, and, along with that 'spin', the attribution of an intrinsic magnetic moment exactly equal to that which an electron circulating in a lowest Bohr orbit would produce electrodynamically: $\mu_B = eh/4\pi m_e c$, the 'Bohr magneton'. (Here e is the electron's charge, m_e its mass, $h/4\pi$ its intrinsic angular momentum, and c the velocity of light.) Since, until neutrons came to the rescue in 1932, atomic nuclei were thought to be composed of protons and electrons, the attribution to electrons of this atomic size magnetic moment, roughly a thousand times greater than that which atomic nuclei had been thought to possess, presented atomic physicists with a perplexing problem.[11]

This issue was further sharpened with Dirac's publication early in 1928 of an equation that, applied to a particle of the electron's charge and mass, 'produced' exactly the previously attributed intrinsic angular momentum and magnetic moment. With the success of Dirac's equation, it came to be widely accepted in the early 1930s that the proton was describable by the same equation, with appropriate charge and mass—leading to the conclusion that the magnitude of the proton's intrinsic magnetic moment, the 'nuclear magneton', must be exactly $\mu_B(m_e/m_p)$, where m_e and m_p are the masses of the electron and the proton, standing in the ratio of 1 to 1850.[12]

Measurement of the magnetic moment of a nucleus was Stern's cynosure, the particular goal towards which his refinements of the molecular beam technique had pointed from the early 1920s to the early 1930s.[13] The implication of an atom-size nuclear magnetic moment, however perplexing, was enticing to the experimenter. When Rabi, itinerant postdoc in Europe, committed himself to a year's work in Stern's laboratory late in 1927, he was paired with Stern's other American postdoctoral fellow, John B. Taylor. This very skilled experimentalist, to whose example and instruction Rabi would be greatly beholden, was then, at Stern's behest, searching for the large nuclear magnetic moment implied by the spinning electron.[14]

Through the strong support of George B. Pegram, the perpetual head of Columbia's physics department and occupant of various higher administrative positions in the university, in the autumn of 1929 Rabi came back to Columbia as faculty member. Although research funds at Columbia were not especially ample, Pegram ensured that Rabi always had an exceptionally large share of them, as well as an exceptionally large share of the time of the Physics Department's exceptionally well-equipped machine shop. Moreover, in the enormous, 14-storey, physics building that Columbia had completed as Rabi was completing his doctoral research in 1926, there was ample space through the 1930s for Rabi's ever-expanding research group.[15]

Rabi's return to Columbia in the summer of 1929 coincided with Gregory Breit's arrival in New York to begin teaching at New York University—indeed, to take, at twice Rabi's salary, intellectual leadership of physics at NYU. Breit was Rabi's age, but was his senior in every professional sense, having been precocious rather than backward in finding his way into theoretical physics. Industrious and learned, he already had a huge record of publication, and he did much of his work in close collaboration with experimentalists. His particular concern at this time and on through the early 1930s was the calculation of nuclear magnetic moments and of hyperfine splittings, and comparison of the results with spectroscopic observations.[16]

Soon after their simultaneous arrivals, Rabi and Breit initiated a joint Columbia-NYU seminar in theoretical physics. Thus it was inevitable that Breit and Rabi would come to discuss the problem of atomic beam measurement of nuclear magnetic moments. Rabi was well aware of the difficulties of such measurements after his year in Stern's laboratory: they required a precise 'mapping' of the magnetic field and its gradient in that extremely narrow channel through which the beam passed between the poles of the magnet (Figure 2), while the very smallness of nuclear magnetic moments required the highest possible gradients, and hence the narrowest possible channels. The prospects for such measurements, and especially for such as could compete with those by optical spectroscopists studying hyperfine structures, would not have looked good to Rabi—and all the less good as he well knew that his own strength lay in conceptual tricks, not in refined, precise experimental technique.

Though it was presumably here, with *magnetic* moments, that Breit and Rabi's cogitations over molecular beam measurements of nuclear moments began, it was to *mechanical* moments that they led. Contrary to what is usually stated, the remarkable conceptual trick upon which they came was not a method for measuring nuclear magnetic moments. Rather, the essence of the Breit-Rabi method, clearly stated in the title of their paper,[17] was to ignore the extremely difficult task of quantitatively measuring nuclear magnetic moment, and concentrate rather upon the far less demanding, merely semi-quantitative task of evaluating the integer, or half-integer, quantum number determining the angular momentum (mechanical moment, 'spin') of a nucleus.

Breit and Rabi pointed out that under certain experimental conditions the relatively large electronic magnetic moment of an atom could serve as a 'handle' on the angular momentum vector of the nucleus. The condition therefor was that the magnetic fields used to deflect the atoms must remain so weak as not to disrupt the coupling between the angular momentum of the nucleus and the angular momentum of the extranuclear electron cloud.

The criterion developed by Breit and Rabi for the maintenance of this coupling was that the ratio $g\mu_B H/\Delta W$ be much less than 1. Here H is the applied magnetic field; μ_B is, as before, the Bohr magneton; g is a known number, generally between one and two, characterizing the way in which the angular momenta of the extra-nuclear electrons are themselves coupled together in the particular atomic state considered; and ΔW is the width of the hyperfine structure in the optical spectrum of the atom (expressed in energy units, $h\Delta\nu$). Physically, this denominator ΔW is the energy of the (undetermined) nuclear magnetic moment, μ_I, in the magnetic field produced by the extranuclear electron cloud, H_J, while the numerator $g\mu_B H$ is the energy of that cloud in the applied magnetic field H. And the requirement that the former be much greater than the latter is equivalent to requiring that the torque exerted by the external magnetic field on the electron cloud be much less than that exerted by the electron cloud on the nucleus, i.e., that the mechanical coupling between the angular momentum of the nucleus and that of the electron cloud not be disrupted by the applied magnetic field.

Breit and Rabi pointed out that the preservation of this coupling results in the splitting of the atomic beam into a multiplicity of beamlets whose intensities and separations are not greatly different and whose number, $(2J + 1)(2I + 1)$, yields the number of quantum units of nuclear angular momentum. (Here J is the known angular momentum quantum number of the extra-nuclear electrons, and I is the unknown number of quantum units of angular momentum of the nucleus.) When the condition for maintenance of this coupling was fulfilled, evaluating nuclear spin would merely require determining with certainty the number of 'beamlets' into which the primary beam was split in the magnetic field. No precise knowledge of the strength or gradient of the field and no quantitative measurement of the deflection of the 'beamlets' was required.

This concept, and with it the limitation to the determination of nuclear spin, had obvious appeal, but its implementation was by no means unproblematic. In particular, the condition for the application of the Breit-Rabi method was abandonment of the strong magnetic fields (and field gradients) that had always been employed in magnetic deflection experiments, for the greater the field and gradient, the greater the beam deflection, and hence the greater the sensitivity and precision of the experiment. Now, weak fields had to suffice.

Early in 1931 Rabi began to build up a molecular beam apparatus 'on an American scale' (as Rabi wrote Stern) to demonstrate the practicability of the Breit-Rabi method. In order to achieve perceptible separations of the 'beamlets' with the weak magnetic fields and field gradients required by the Breit-Rabi method, Rabi's design provided a beam path of almost 40 cm, twice the lengths Stern had used. Otherwise, however, the apparatus was a rather crude replica of John Taylor's Hamburg apparatus. The results that Rabi reported at the end of 1931 could be construed as evidence for the existence of nuclear spin only with some good will. For an evaluation of that spin his results were entirely insufficient.[18]

Rabi's only hope for doing better was to find collaborators gifted for experimental work. Late in 1931 he got one in graduate student Victor William ('Bill') Cohen, and in the autumn of 1932 he was able to hire a postdoctoral research assistant, Carl Frische—for one year only.

Thereafter the number of skilled experimental collaborators—students and, especially, postdocs—grew steadily: Sidney Millman, Jerome Kellogg, Jerrold Zacharias, John Manley, ...[19]

Meanwhile, back in Hamburg, through 1932 and during the first half of 1933—until forced to emigrate in summer 1933—Stern used his high-gradient technique in several very difficult experiments on hydrogen molecules and deuterium molecules in order to measure the intrinsic magnetic moment of the proton to within 10% and make a rough estimate of the magnetic moment of the neutron. He found that μ_p was not the the nuclear magneton, $\mu_B(m_e/m_p)$, that physicists, following Dirac's 1928 theory of the electron, confidently expected it to be, but some $2^1/_2$ times this value.[20]

Stern's result being, without question, the most important relating to nuclear moments that had yet come to light, it cried out for confirmation by a different method. The Breit-Rabi 'indirect' method was an alternative—the only alternative—but only if Rabi could extend it from determination of nuclear spin to measurement of nuclear magnetic moment. This was indeed possible in the case of hydrogen atoms, because for one-electron atoms, and only those, it was possible to calculate numerically, exactly, the magnetic field H_J produced at the nucleus by the 'electron cloud'. To take advantage of this possibility, however, more was required from a Breit-Rabi experiment, namely, just what was required in a traditional Stern-Gerlach experiment: precise knowledge of the deflections produced and of the strength and the gradient of the magnetic field producing them.

Rabi's whole prior career was based on circumventing field gradient measurements, and he was not disposed to accept the necessity of such now—especially now, where the weak fields employed in the Breit-Rabi method tended rather to increase the significance of measurement uncertainties. The alternative to measuring was calculating: to produce a magnetic field by a means that permitted an accurate and precise calculation of its strength and gradient. Such were the fields produced by electric currents in the absence of ferromagnetic materials. Fields so produced could never be very strong, and therefore had not previously been employed in atomic beam magnetic deflection experiments. But the very point of the Breit-Rabi method was to remain in a regime of weak fields.

Rabi found that two parallel 'wires' (in practice, copper tubes) with electric current flowing in them in opposite directions (and water as coolant flowing through them) produced a magnetic field that was nearly constant in planes parallel to the plane through the two wires (center lines of the copper tubes). And as the magnetic field was constant in such planes, so also was its gradient. In these planes the ribbon-like atomic beam would lie, and all atoms within the beam thus would be subject to the same deflecting field. This allowed still broader ribbons, and consequently higher beam intensities, than had ever been possible with the wedge-and-groove Stern Gerlach magnets. With such a deflecting magnet—this first was only 15 cm long—Rabi and collaborators determined the proton magnetic moment in the spring of 1934, confirming the anomaly, and indeed finding it even larger than had Stern.[21]

Late in 1933 Rabi designed and had Columbia's skilled machinists construct for doctoral student Sidney Millman—who was gearing up to follow up Cohen's wedge-and-groove measurements on alkalis—a 61.5 cm long version of the '2-wire' magnet (Figure 3). This, with great certainty, is the 2-foot magnet here denominated the 'Millman' magnet.

Having found that the 2-wire scheme worked so well, Rabi naturally wanted to see whether the uniformity of the magnetic field and the strength of its gradient could be improved by more complicated configurations of parallel wires. The most obvious refinement was the addition of a second pair of parallel wires, their centers forming a square with those of the first pair, but powered independently. This indeed is the design of our 2-foot magnet (the 'Millman' magnet). Occasionally, but only occasionally, employing the second pair of wires, Millman was able to gain some advantage in his experiments on potassium-39.[22]

Millman's magnet continued in service in Rabi's laboratory, first in experiments by Marvin Fox, Millman's understudy, on the nuclear

Figure 3. The 'Millman' and the 'Manley' magnets as depicted in publications from Rabi's laboratory: Figs 3a and 3b, 'Millman' magnet in 1935 and in 1937;[22, 23] Figure 3c, 'Manley' magnet in 1936.[24]

In Figure 3a the 'Millman' magnet, marked 'E', appears at the center of the apparatus. The four copper tubes ('FIELD WIRES'), separated only by thin sheet of micas, lie against each other, embedded in the face of the aluminum alloy bar ('field block'). Here, as also in Figs 3b and 3c, as indeed was then standard in magnetic deflection experiments, the ribbon-like beam is vertical (stands on edge), and passes from oven 'A' to detector 'G' just 'in front' of the field wires. That there were originally two sets of field wires is clearly indicated by the two pairs of lines diverging from the right end of the field block to either side of the collimating slit 'D'.

In Figure 3b the 'Millman' magnet, marked 'B', has been reused by Torrey in a rather more complicated experiment.[23] Here the orientation of the magnet is unchanged, but the oven and detector positions are switched. Thus the collimator slit 'D', with its brass frame bolted to the duralumin field block, is now at the 'trailing', not the 'leading' end of the magnet. Here only two field wires are shown at the right end of the magnet, indicating the desuetude, if not already the removal, of the second pair of tubes by 1937.

Figure 3c shows the apparatus, essentially similar to that of Figure 3a, in which the lengthy 'Manley' magnet, marked 'D', was first used. Here the two pairs of tubes feeding into the field block at its center are clearly indicated as tubes rather than schematically as wires.

All three drawings depict, in progressively increasing detail, a contact ionization detector 'G'. Developed by John B. Taylor in 1928, the year Rabi shared his Hamburg laboratory, this was the first electronic detector of molecular or atomic beams, as also the first highly sensitive and truely quantitative detector. It was indispensable for all of Rabi's experiments at Columbia employing alkali atoms, i.e., the overwhelming majority of his experiments. At its center is a fine tungsten wire parallel to the plane of the beam and translatable perpendicular to that plane. The tungsten wire is held at a high temperature and at a moderate positive potential, and is surrounded by a cylindrical cage connected through a sensitive galvanometer to a negative potential. Alkali atoms, on striking the hot tungsten, surrender their one, loosely bound, valence electron to it, and are then repelled to the surrounding cage. The current through the galvanameter so produced is thus a precise measure of the atom flux at the position of the wire.

SIDNEY MILLMAN

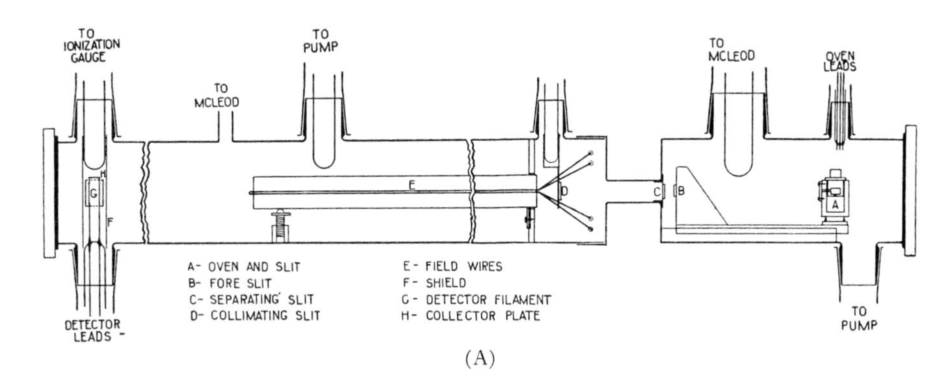

(a)

MAGNETIC MOMENT OF POTASSIUM NUCLEUS

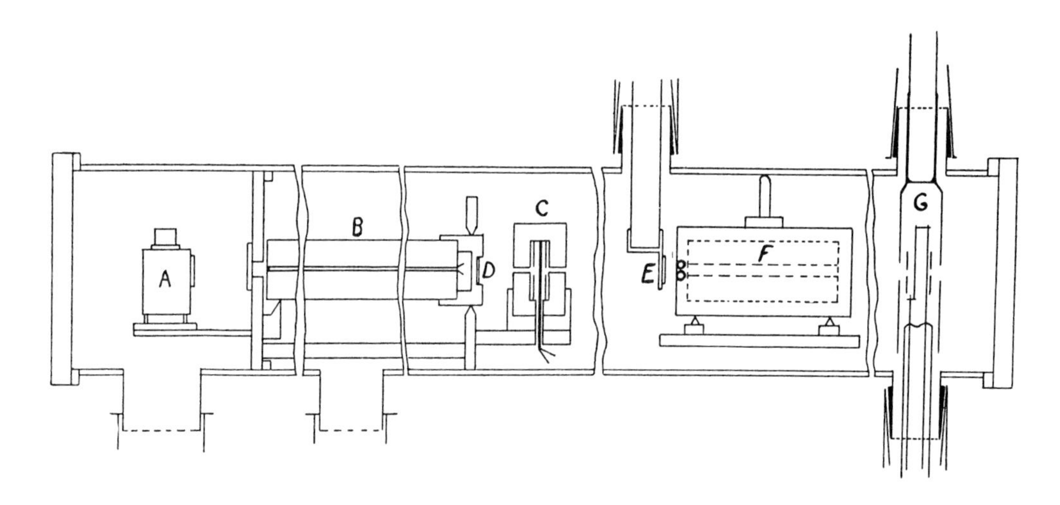

(b)

J. H. MANLEY

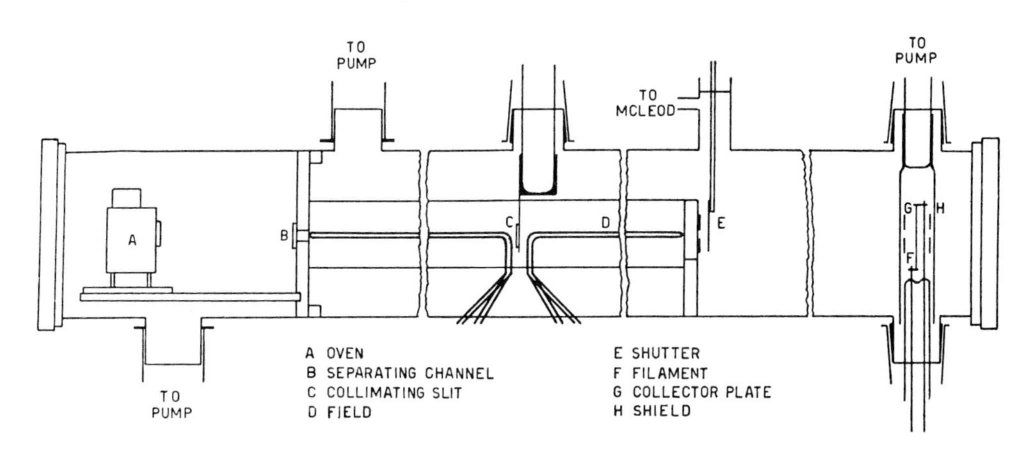

(c)

moments of potassium-41, and then by graduate student Henry Torrey, 1936–37, in an experiment to determine the sign of the potassium-39 nuclear magnetic moment, i.e. whether the magnetic moment is directed parallel or anti-parallel to the nuclear spin. It appears that the form in which the 'Millman' magnet has come down to us is that in which it was left by Torrey, with his brass frame bearing an adjustable-slit attached to its right end.[23]

John Manley, with a doctorate from the University of Michigan, arrived at Columbia late in the summer of 1934 to take up a sub-faculty position as Instructor. Looking about to see who had interesting research in progress, he came to Rabi, who really had little competition in that regard. Manley's background included a B.S. in 'engineering physics', and perhaps for that reason Rabi set him to work on an atomic beam apparatus (Figure 3) that made Rabi's original 'American scale' Breit-Rabi apparatus appear Lilliputian. The nucleus in question was, once again, potassium 41, which because of its very small nuclear magnetic moment and low abundance, had resisted Millman's and Fox's (and Rabi's) efforts to determine its spin with certainty. And when Manley had succeeded well there, Rabi had him proceed to lithium 6 and 7 with the same apparatus.[24]

The apparatus (Figure 3c), more than two meters long, contained a deflecting magnet of length 153 cm, that is, again with considerable certainty, our 5-foot 'Manley' magnet. Although constructed so that the tubes in each half of the bar are fed separately at its center, it was operated integrally—two feeds being provided only to reduce electrical resistance and facilitate cooling. Further, although the published schematic drawing of the apparatus shows it to have been constructed as a '4-wire' magnet, as indeed ours is, neither Manley, nor John Gorham who used it after him, reported having used it otherwise than as a 2-wire magnet.

The Breit-Rabi method was a brilliantly direct method for the determination of nuclear spin, but it was an only indirect, and in most cases insufficient, method for measurement of nuclear magnetic moment. Stern took hydrogen molecules with zero electronic angular momentum and zero electronic magnetic moment, and operated directly upon their very small nuclear magnetic moment, μ_p, by means of highly inhomogeneous magnetic fields. Consequently, the deflections that he measured were (with appropriate corrections) direct measures of μ_p. Rabi, on the contrary, required atoms with non-zero electronic magnetic moment, for it was that thousand-times larger electronic magnetic moment upon which the Breit-Rabi method relied to get observable deflections with weak magnetic fields. And he could put a number on the nuclear magnetic moment, μ_I, only when theory could provide him with an estimate of H_J, the magnetic field at the nucleus due to the electron cloud.[25]

Only for hydrogen (i.e., one-electron atoms) could theory provide a precise H_J, and only for the alkalis (i.e., atoms with only one valence electron) could it provide even an approximate value. Such a situation is never entirely congenial to the experimentalist, and it was especially uncomfortable in the mid-1930s as optical hyperfine spectroscopy was producing very different values of μ_I, depending upon which hyperfine structure—i.e., which electronic state—was used to evaluate μ_I.[26] Since Rabi was limited to the evaluation of μ_I in just one state—the ground state—and that only where theory had an H_J to give him, he was hard pressed to make any very strong claim for his number.

It was thus the 'logical conclusion' of the Breit-Rabi method to extend it in such a way as to wrest from it a direct measurement of μ_I. Indeed, the possibility had been implicit in the Breit-Rabi calculation all along, had it not treated μ_I as negligibly small compared with μ_B (i.e., compared with μ_J, the electronic magnetic moment). In 1936 Rabi reconsidered the calculation from that point of view, and found that carrying μ_I through implied a more complicated beam splitting. In the mode of Breit-Rabi measurement introduced in 1934—the so-called 'zero moment' method—treating μ_I/μ_J as small but non-negligible implied that the peaks in detector current were actually narrow doublets. (In the 'zero moment' method, rather than mapping the pattern of beamlets at a fixed magnetic field, the detector was fixed at the position of the undeflected beam, and the 'pattern' was swept over the detector by gradually increasing the current through the magnet.) The small difference in magnetic field ΔH separating these two maxima in detector current is proportional to H and to μ_I/μ_J.

If the Breit-Rabi technique could be made sensitive enough to make this fine structure perceptible, μ_I would be obtained free of that factor for which Rabi was dependant upon theoretical calculation, as also free of the assumptions about the physical nature of the interaction between the magnetic moment of the nucleus and the extranuclear electrons on which the hyperfine structure determinations depended. With such a measurement Rabi would have obtained the 'absolute nuclear moment.'[27] To carry this through, Rabi needed a nucleus with a large magnetic moment in an atom with small (but non-zero) μ_J and a readily detachable valence electron (so as to behave like an alkali atom in a contact-ionization detector).

Rabi fixed on indium. Its large nuclear magnetic moment assured the satisfaction of the Breit-Rabi coupling condition at relatively high magnetic fields—10,000 gauss. Such fields, however, were larger than could be obtained without the aid of iron. Rabi therefore designed an iron-core magnet with pole faces that would reproduce the field of a 2-wire magnet. The requisite geometry of the pole cross-section proved exceptionally simple: semicircles. A magnet one meter long of Armco iron, accurately milled with poles of approximately 3 mm (1/8 inch) radius of curvature (corresponding to two parallel wires 1/4 inch apart), wound with four turns of copper tubing, was fabricated under hand and eye of

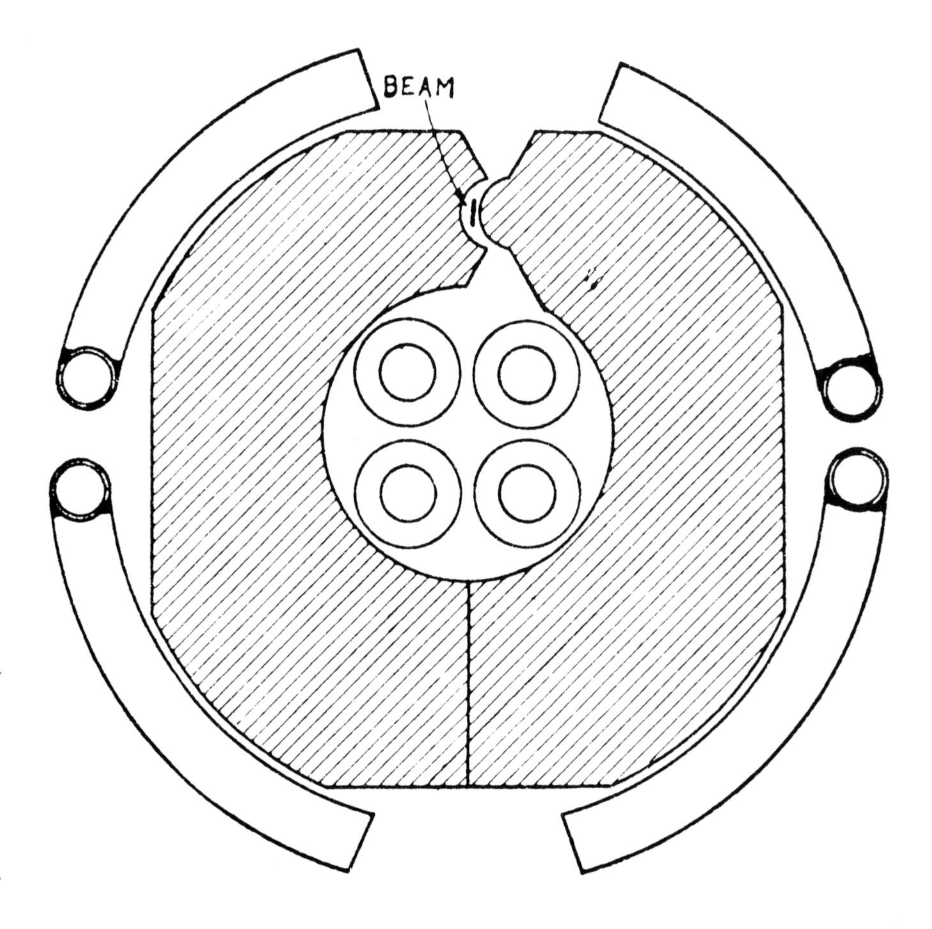

Figure 4. The 'indium' magnet. Figure 4a is 'Cross section of magnet and windings' as depicted in Rabi and collaborators' report on their 1937 experiments.[27] The iron portion of the magnet is shown hatched. The copper windings take the form of thick-walled tubes, running the full meter length of the magnet from end to end within the central hole, and outside the magnet body have the form of segments of hollow copper cylinders (with cooling water carried in thin-walled copper tubes soldered to them). As Rabi et al. state the drawing to be '2/3 full size,' the 17 mm diameter of the central hole as there depicted corresponds closely to the 1 inch measured diameter of the central hole of our 'indium' magnet. (Thus Rabi's drawing is here reproduced '1 1/2 full size'.)

the Department's chief mechanician, Sam Cooey. The apparatus was otherwise essentially similar to that in Fig. 3c.

It is my surmise that the third of our artefacts, the iron magnet with central hole of 1-inch diameter, and of 'length' 1-inch, was sliced off that original 1-meter long magnet as souvenir paperweight—the hinge being then added to hold the two halves together while displaying the finely machined pole faces.[28]

In a series of very difficult experiments, together with Sidney Millman and Jerrold Zacharias, in the summer of 1937 ('the indium summer' Zacharias punned) Rabi carried through this 'absolute' measurement. It was indeed a tour de force, the capstone of the Breit-Rabi research program. Further in this direction it was hardly possible to proceed. In September Cornelius Gorter visited Columbia and urged Rabi to try magnetic resonance.[29] Rabi was ready.

Notes

1. I. I. Rabi, Julian S. Schwinger, Norman F. Ramsey, Sidney Millman, and Jerrold Zacharias, with Jack S. Goldstein, moderator, 'Reminiscences of the thirties,' recollections and discussions videotaped at Brandeis University, 16 March 1984, transcript, 57pp., and videotapes in Zacharias Papers, MIT Archives, on p. 51 of the transcript.

2. Gregory Breit and I. I. Rabi, 'Measurement of Nuclear Spin,' *Physical Review* 38 (1931), 2082–83 .
3. Paul Forman, 'Molecular Beam Measurements of Nuclear Moments before Magnetic Resonance: I. I. Rabi and Deflecting Magnets to 1938. Part I,' *Annals of Science* 55 (1998), 111–160. Parts II and III to appear in future.
4. Rabi to F. K. Richtmyer, 22 January 1929, draft (Library of Congress, Rabi Papers, box 7, folder 7).
5. The most complete collection of references to pre-World War II molecular beam research is W. H. Bessey and O. C. Simpson, 'Recent Work in Molecular Beams,' *Chemical Reviews* 30 (1942), 239–279. For the magnetic resonance technique, as also for valuable discussions and references on earlier techniques: Norman F. Ramsey, *Molecular Beams* (Oxford, 1956; reprinted 1990).
6. This circumstance is clearly 'illustrated' in photographs taken of Rabi at about the time he was himself giving up active research work—photographs, reproduced in the introduction to the *Annals of Science* paper cited above (note 3), in which Rabi chose to have himself shown holding one or another of these artefacts. Even decades later when Rabi had long forgotten the existence of these artefacts buried in the clutter of his office, *this* remained the work that he wanted to talk about when he had the ear of a historian of physics. Thus in 1984 S.S. Schweber sought to interview Rabi about the work at Columbia after World War II bearing upon quantum electrodynamics. Rabi, however, turned and returned the interview to his pre-magnetic-resonance experiments and their fundamental importance. Rabi, interview, 1984 February 13, by S.S. Schweber, transcript, 25pp., in Niels Bohr Library, American Institute of Physics.
7. Papers and data of Angelo Baracca and collaborators as cited and reproduced in Forman, '"Swords into Ploughshares": Breaking New Ground with Radar Hardware and Technique in Physical Research after World War II,' *Reviews of Modern Physics* 67 (1995), 397–455, on pp. 420–21. J. L. Heilbron and R. W. Seidel, *Lawrence and his Laboratory: A History ...* (Berkeley, 1989).
8. Forman, '"Swords into Ploughshares"....,' p. 405. Rabi's technique—the Breit-Rabi method—was really only *semi*-independent. Only in the 1937 experiment employing the 'indium' magnet did Rabi & Co. succeed in a measurement of nuclear magnetic moment that did not depend upon elaborate and sometimes uncertain theoretical calculations. Alone and of itself, all that a Breit-Rabi experiment can provide is a rough, rather qualitative, estimate of the magnitude of the magnetic moment of the nucleus, based on observation of the range of applied magnetic field strengths in which decoupling of the electronic from the nuclear angular momentum is first produced.
9. K.K. Darrow to Rabi, from San Francisco, 6 August 1936 (Library of Congress, Rabi Papers, box 2, fldr 11). Rabi had presented his work at the Cornell Symposium on Nuclear Physics early in July, and had lectured at the Ann Arbor Summer School during the last two weeks of July.
10. Kronig, 'Spinning Electrons and the Structure of Spectra,' *Nature* 117 (1926), 550. Rabi, 'Spinning Electrons,' *Nature* 118 (1926), 228. Forman, 'Molecular Beam Measurements ... Part I,' pp. 137, 142–43. Pauli left Hamburg for Zurich in spring 1928, where Rabi rejoined him early in 1929.
11. Roger H. Stuewer, 'The Nuclear Electron Hypothesis,' in William R. Shea (ed.), *Otto Hahn and the Rise of Nuclear Physics* (Dordrecht, 1983), pp. 19–67.
12. Forman, 'Molecular Beam Measurements ... Part I,' p. 145; Helge S. Kragh, *Dirac: A Scientific Biography* (Cambridge, 1990).
13. Stern, 'Zur Methode der Molekularstrahlen,' *Zeitschrift für Physik* 39 (1926), 759–60. Stern himself was the first to introduce the concept of a 'nuclear magneton,' pointing out that general arguments implied that if the atomic nucleus is a dynamical system containing proton-mass particles with electron-size charges and with angular momenta quantized in the same units as that of the extra-nuclear electrons, then the magnetic moments of nuclei should be roughly two thousand times smaller than those of atoms, i.e., should be of the order of $\mu_B(m_e/m_p)$.
14. John B. Taylor, 'Das magnetische Moment des Lithiumatoms,' *Zeitschrift für Physik* 52 (1928), 846–52. Forman, 'Molecular Beam Measurements ... Part I,' p. 127–9, 144–58.
15. Columbia University, *The New Physics Laboratories of Columbia University in the City of New York*, 1927 (Privately printed: New York, 1927). L. A. Embrey, 'George Braxton Pegram,' National Academy of Sciences of the USA, *Biographical Memoirs* 41 (1970), 357–407. I. I. Rabi, 42 interviews for the Columbia University Oral History Research Office, 1983–85, by Chauncey Olinger, 1102pp. continuously paginated transcript, on pp. 233–34, et passim.
16. John Archibald Wheeler with Kenneth Ford, *Geons, Black Holes, and Quantum Foam: A Life in Physics* (New York, 1998), Chapter 5 ('I Try My Wings [with Breit]'), pp. 103–123. D. A. Bromley and V. W. Hughes (eds.), *Facets of Physics* [Breit Festschrift] (New York, 1970).
17. Breit and Rabi, 'Measurement...' op. cit., note 2.
18. I. I. Rabi, 'The Nuclear Spin of Caesium by the Method of Molecular Beams,' *Physical Review* 39

(1932), 864. This is a 200-word abstract of a paper presented at the American Physical Society meeting New Orleans, 29–30 December 1931. Details of the development of concept and apparatus are to be found in Rabi's letters to Stern, 3 June 1931 and 18 January 1932, in University of California, Berkeley, Bancroft Library, Stern Papers, carton 1, folder: Rabi.

19. John S. Rigden, *Rabi: Scientist and Citizen* (New York, 1987). Jack S. Goldstein, *A Different Sort of Time: The Life of Jerrold R. Zacharias, Scientist, Engineer, Educator* (Cambridge MA, 1992). Victor W. Cohen, untitled recollections sent to Lucy Hayner, 27 July 1956, 5pp. typescript, in Columbia University Rare Book and Manuscript Library, George B. Pegram papers, box 3, folder: Cohen.
20. Announced in I. Estermann, R. Frisch and O. Stern, 'Magnetic Moment of the Proton,' *Nature* 132 (1933), 169–70, with details provided in the *Zeitschrift für Physik.*
21. Rabi, Kellogg and Zacharias, 'The Magnetic Moment of the Proton,' *Physical Review* 46 (1934), 157–63.
22. Sidney Millman, 'On the Nuclear Spins and Magnetic Moments of the Principal Isotopes of Potassium,' *Physical Review* 47 (1935), fig. 2 on p. 742; Millman, 'Recollections of a Rabi Student of the Early Years in the Molecular Beam Laboratory,' *Transactions of the New York Academy of Sciences* 38 (1977), 87–105.
23. Henry Cutler Torrey, 'The Sign of the Magnetic Moment of the K^{39} Nucleus,' *Physical Review* 51 (1937), fig. 2 on p. 503.
24. J. H. Manley, 'The Nuclear Spin and Magnetic Moment of Potassium (41),' *Physical Review* 49 (1936), fig. 1 on p. 922; Manley and Millman, 'The Nuclear Spin and Magnetic Moment of Li^6,' *Physical Review* 51 (1937), 19–21; John E. Gorham, 'The Signs of the Nuclear Magnetic Moments of Li^6 and K^{41},' *Physical Review* 53 (1938), 563–67.
25. Marvin Fox and I. I. Rabi, 'On the Nuclear Moments of Lithium, Potassium, and Sodium,' *Physical Review* 48 (1935), 747.
26. H. A. Bethe and R. F. Bacher, 'Nuclear Physics,' *Reviews of Modern Physics* 8 (1936), 206–225, reprinted in Hans A. Bethe, et al., *Basic Bethe: Seminal Articles on Nuclear Physics,* 1936–1937 (New York, 1986).
27. S. Millman, I. I. Rabi, and J. R. Zacharias, 'Absolute Nuclear Moment of Indium 115,' *Physical Review* 53 (1938), 331; S. Millman, I. I. Rabi, and J. R. Zacharias, 'On the Nuclear Moments of Indium,' *Physical Review* 53 (1938), 384–91.
28. The stated lengths and gap dimensions of the deflecting magnets used in Rabi's early post-World-War-II molecular beam magnetic resonance experiments on hydrogen—John E. Nafe and Edward B. Nelson, 'The Hyperfine Structure of Hydrogen and Deuterium,' *Physical Review* 73 (1948), 722–3—together with the reference there directly back to the description of the magnet used in Rabi and collaborators's 1937 Breit-Rabi method experiments on indium (op. cit., note 27), suggest that the original indium magnet was cut into shorter lengths for these early (and urgent) postwar experiments, and that Rabi's souvenir was a by-product of this process.

    It is worth noting that the problem of identifying our 'indium' magnet with certainty is greatly aggravated by the circumstance that this basic design, devised for indium in 1937, was then adopted for the several magnets constructed in subsequent years for experiments employing the magnetic resonance technique. Indeed, the iron emulation of the two-wire field by semi-circular pole faces became *the* standard magnet design during the following two decades in which the molecular beam technique spread from two or three centers to two or three hundred.
29. I. I. Rabi, J. R. Zacharias, S. Millman, and P. Kusch, 'A New Method of Measuring Nuclear Magnetic Moment,' *Physical Review* 53 (1938), 318. C. J. Gorter, 'Bad Luck in Attempts to Make Scientific Discoveries,' *Physics Today* (January 1967), 76–81.

Bernard Finn

Collectors and Museums

Somewhere in our genetic structure, perhaps between the code for perfect pitch and the one for a tendency towards overweight, one may imagine lies the gene for collecting. It is one we apparently share with other species—sometimes practical, as with the pack rat; sometimes aesthetically or otherwise driven, as with the blue jay. And it can be a very significant motivating force in our lives.[1]

Indeed, the collecting instinct is often directed in such a way that it is a defining element in our character. Long before settling on a career path, many a naturalist has assembled elaborate collections of birds eggs or leaves or rock samples. Charles Darwin is only one of numerous notable examples. He wrote that 'By the time I went to this day-school my taste for natural history, and more especially for collecting was well developed.' He described it as a 'passion ... [that] was very strong in me, and was clearly innate....'[2] For others, collecting has been a relaxing avocation, independent of regular professional activities. Franklin Roosevelt and stamps, for instance.[3] But what happens to these collections? For many of us, myself included, they end up tucked away in an attic storage box or scattered to grandchildren. (Darwin's ended up in numerous museums, Roosevelt's stamps were separated and sold at auction by his family after his death.)[4] But many people have chosen to institutionalize their passions, thus creating permanent memorials to themselves. If the collector has enough money (or proper connections), he or she may achieve the ultimate goal of having the collection enclosed in a marble shrine—usually called a museum—with his or her name inscribed over the door. Lesser mortals are content with having their names attached to special collections in museums or perhaps to exhibition halls.

The point to be made here is that these collections do more than memorialize their compilers. Collectors help to shape museums, and therefore to shape the ways we present our cultural identities to ourselves and to others. Sometimes this is very conscious. Examples from the United States include Rolf Klep, who made his collection of maritime artefacts and artwork the centerpiece of a new Columbia River Maritime Museum in 1962, for the expressed purpose of drawing attention to the maritime heritage of the Northwest United States. With additional funds and objects from other sources this has become a successful museum that in 1991 was designated by the Oregon legislature as the official maritime museum of the state.[5] On the other side of the country, George Hewitt

Meyers and his wife collected non-European rugs and textiles. These became the focus of the Textile Museum in Washington in 1925, which in turn has become a center for the study of textile traditions of non-Western cultures.[6] Still another example is Earl Bakken who founded The Bakken: A Library and Museum of Electricity in Life in 1976 on the basis of his collection of electro-medical devices. His purpose was 'to further understanding of the history and applications of electromagnetism in the life sciences and to benefit contemporary society.'[7] In 1949 Etta Cone left the extensive collections of twentieth-century art that she and her sister Clarabel had accumulated to the city of Baltimore for the purpose of 'improving the spirit of appreciation of modern art in Baltimore.'[8] Whatever effect this may have had on the citizens of their city, the gift has had a defining impact on the Baltimore Museum of Art which has made the Cone bequest a cornerstone for continuing collecting efforts in twentieth-century art. The Museum of International Folk Life in Santa Fe, New Mexico, which opened in 1953, was based on the extensive collections of Florence Dibell Bartlett. She composed the museum's motto, 'The art of the craftsman is a bond between the peoples of the world.' [9] The extent of the collections, but not the scope, was increased some five-fold by the addition of the collections of Mr. and Mrs. Alexander Girard in 1978. [10] It is significant that both gifts stimulated substantial commitments by the state.

Often the transformation from private collection to public institution has been less self-conscious, with consequences that have been less predictable. Perhaps the most notable example is that of Hans Sloane (1660–1753), whose position as a physician to the social elite in London provided him with both income and contacts that would be important in exercising his collecting instincts. Eighteen months in Jamaica (1687–89) provided his first major opportunity to obtain a substantial number of natural history specimens, and on his return to London he quickly expanded his scope to include an encyclopedic range of man-made artefacts, ancient and modern. Over the years these were augmented by museum-size acquisitions from other collectors. When Sloane died, his will provided that all of these collections should be offered to the government for £20,000, to be paid to his daughters. After some efforts on the part of Sloane's trustees, Parliament agreed to the terms. This was to be arranged at no expense to the Treasury through the mechanism of a public lottery—a common eighteenth-century expedient—which in this case netted the new museum a little over £95,000. (It was also a particularly corrupt lottery, leading to the general abandonment of the process until its revival in the late twentieth century, when, one might say fittingly, the British Museum and other cultural institutions have again become beneficiaries.)[11]

In New York, across the Atlantic and across two centuries, on a much smaller scale, the taxidermy collections of O. Carol Lempfert were the stimulus for The Suffolk Museum of Stony Brook, chartered in 1942 as 'an association to increase and diffuse knowledge and appreciation of history, art and science, ... to protect historic sites, works of art, scenic places and wildlife from needless destruction, to provide facilities for research and publications, and to offer popular instruction and opportunities for aesthetic enjoyment...'[12]

Of particular interest, however, is a gift that reached from one side of the ocean to the other. Under the terms of James Smithson's will, in the event that his nephew should die without issue (as happened in 1835), his estate was left to the United States Government 'for the increase and diffusion of knowledge.' As is well known, the cash proceeds, which amounted to approximately $500,000, were eventually placed in an endowment which funded the construction of a building (the present Smithsonian Castle) and paid for operating expenses in the early years. But there were also many mineral specimens, which were placed with other government collections in the Patent Office until 1857. In that year the Congress authorized expenditure of funds for the housing of these collections by the Smithsonian, augmenting a limited museum of natural history that had already been formed within its walls. Thus, although it is incorrect to suggest that Smithson's minerals were the cornerstone of the present museum enterprise in Washington, they were a significant part of his legacy which eventually led to this wide-ranging complex—the extent of which is far beyond anything that James Smithson could have imagined.

Other collectors have made their mark on the Smithsonian. Several (all art collectors) fit in the category, mentioned above, of individuals who were able to parlay their collections into full-blown museums. These include the (Charles Lang) Freer Gallery of Art in 1923, the (Joseph H.) Hirshhorn Museum and Sculpture Garden in 1974, the Arthur M. Sackler Gallery in 1987, the (Warren Robbins) National Museum of African Art in 1987, and the George Gustav Heye Center (in New York City) of the National Museum of the American Indian in 1994. Some, like the Freer in particular, have had very restricted collecting policies; others have been able to augment their holdings to a significant degree. But all have kept quite close to the scope envisioned by their founders. The result is an eclectic mix of art museums, the product, one might say, of 'targets of opportunity' that have appeared over the years.

One might add to this list the (Andrew Mellon) National Gallery of Art which opened to the public in 1941, and which has a loose administrative affiliation with the Smithsonian. Of special interest here is that Mellon financed a building that was much too large for his relatively modest collection of paintings, which was broad in scope and high in quality but numbered 115 (not counting American portraits). Other gifts were anticipated, and even before the museum opened Mellon's paintings were augmented by collections from Samuel Kress and Chester Dale, and soon after from Joseph Widener, Lessing Rosenwald, and others. This gave the museum an extraordinarily fast start in its development as a world class institution.[13]

Indeed, it is the role of most donors to present their offerings to existing institutions. But this doesn't prevent them from shaping the directions those museums take. The Museum of American History is no exception, and thousands of collectors have helped to nudge it along one course or another. Let me take one area, electricity, as an example.

In 1847 the Smithsonian's Board of Regents determined 'That it is the intention of the act of Congress, and in accordance with the design of Mr. Smithson, that one of the principal modes of executing the act and the trust, is the accumulation of collections of specimens and objects of natural history and of elegant art...'[14] Joseph Henry, Secretary of the Institution, was fundamentally opposed to a major commitment in this direction, fearing that concern for the care of collections would overwhelm his limited resources. But he chose naturalist Spencer Baird as his assistant secretary in 1850 and Baird, like virtually everyone in his discipline, had a large personal collection of natural history specimens. He brought them to Washington and continued to collect on an increasingly large scale. The term 'museum' began to be used in annual reports to describe these holdings as early as 1850. Henry succumbed officially in 1857, when Congress gave the Institution custody of several important government collections, together with $4,000 per year to care for them. The museum took physical form in 1881, when the new Arts & Industries Building opened its doors.

The new museum continued to be dominated by natural history specimens. Visitor's guides in the mid-1880s described a modest number of relics of presidents (among other items, clothing from Washington, swords from Grant, a lock of hair from Lincoln), a printing press reputedly once used by Franklin, chemical apparatus from Joseph Priestley, and a miscellany of other 'historical' items. (It should be noted that there were also several boats and ship models in the Fisheries section.) But in the 1890s, under G. Brown Goode—a naturalist who had wide-ranging interests—the doors were literally opened to historical artefacts of all sorts. To achieve his goal of expanding the museum's mission, he quickly established a substantial curatorial staff—mostly with honorary, unpaid enthusiasts. It was these enthusiasts, many of them collectors—H. G. Beyer for Material Medica, J. E. Watkins for Transportation, S. R. Koehler for Graphic Arts—who determined the course that the museum would follow.[15]

One of Goode's curators was George Maynard. A telegrapher as a young man (which included a period in Lincoln's White House at the end of the Civil War), he was an entrepreneur who took advantage of the emerging electrical industry, selling equipment and installing telegraph and telephone lines in Washington. He knew or at least had business relations with a number of the American pioneers like Alexander Graham Bell, Thomas Edison, and William Wallace. He also developed an interest in history, and in 1881 he joined the telegraphers' Old Timers Association, becoming its president ten years later. In 1892 he was named historian of the group and started a campaign to encourage collecting old apparatus. The result was a new organization, the Telegraph Historical Society of North America (1894) which under Maynard's guidance assembled a substantial number of early instruments, many of which formed the core of the Smithsonian's early holdings in this area.[16] (For information about Maynard I am indebted to Richard Loomis who has shown me a copy of his unpublished biography of Maynard; other information appears in the Smithsonian Annual Reports for this period and from accession records.)

Maynard brought his collecting instincts to the Smithsonian in the spring of 1896 as honorary custodian of the electrical collections; a year later he began a two-decade career as a paid staff member. From this position he used his old network of friends and electrical colleagues to assemble artefacts related to the birth and growth of electrical communications and power. He acquired motors and generators representing the pioneering work of William Wallace, Elihu Thomson, Edwin Houston and Moses Farmer; and incandescent lamps tracing the inventions of Thomas Edison. In 1898 he obtained from Alexander Graham Bell some of Bell's experimental apparatus together with telephones associated with several of his chief competitors (assembled for patent litigation), and in 1901 the museum received a significant donation of items from the widow of Bell's chief rival, Elisha Gray.

In August 1903 Maynard succeeded to the position of curator of the technology collections, leaving no one specifically in charge of electricity. Nevertheless, he continued to influence collecting policy. In 1908 he acquired additional material from Bell; and that same year, when the Patent Office offered the Smithsonian the pick of its collection, it was Maynard who selected a number of important electrical models.

This remained the situation until Maynard's death in 1919. After that there was no one to steer the electrical collecting efforts, and with only one or two exceptions they remained moribund until the 1950s. At that point, in preparation for the new Museum of History and Technology (now National Museum of American History) the curatorial staff was increased by a factor of four, and the significance of those early collecting efforts became apparent. A separate section of electricity was re-established, covering communications and power as well as what might be called electrical science. This arrangement is quite different from what one sees in other museums and has clearly affected the way that collections have subsequently been developed and exhibited.

With one curator in charge of this wide a range of science and technology it is arguably true that collecting efforts have depended upon private

Figure 1. A telegraph key from Maynard's Telegraphic Historical Society.

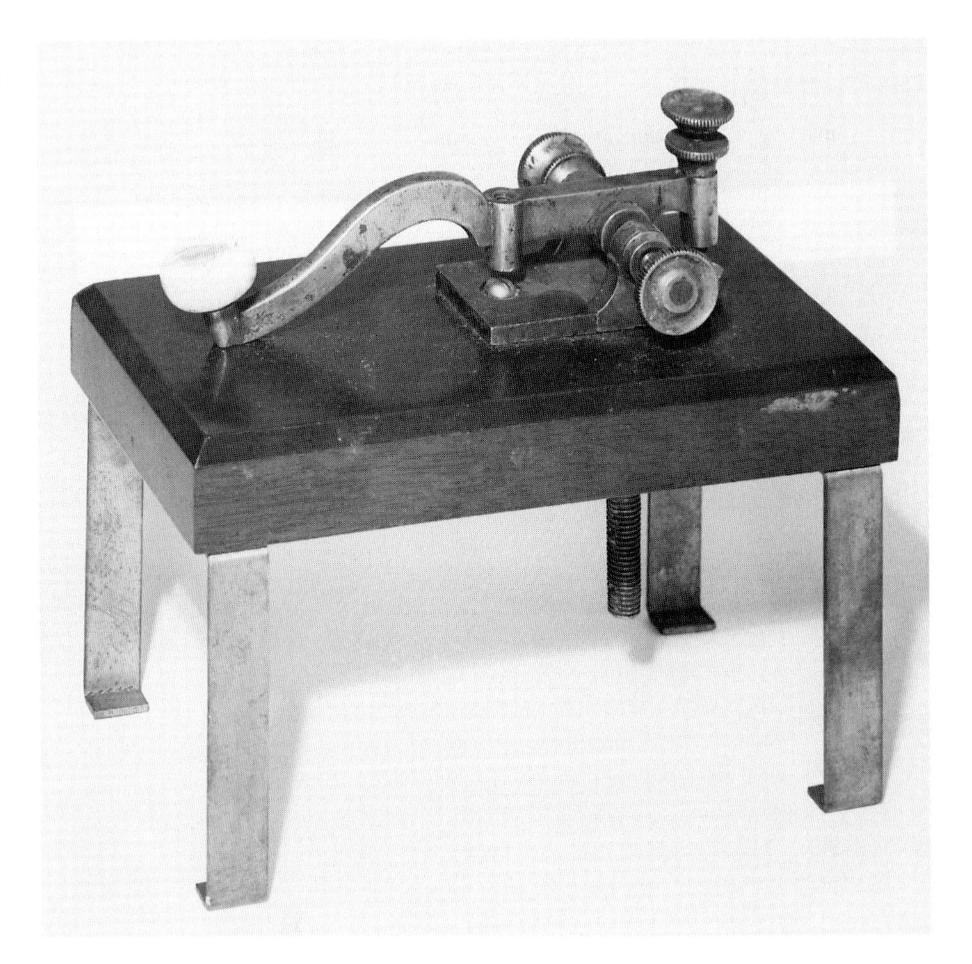

Figure 2. A model of Edison printing telegraph, selected by Maynard from the Patent Office collection in 1908.

Figure 3. One of a number of radio receivers from L. C. F. Horle (actually given after his death by his wife, Susan).

Figure 4. Paul Watson contributed some 1300 vacuum tubes, but also the cases he had constructed to house them.

collectors more than might otherwise have happened. Whatever the case, collectors have been of considerable importance. In giving us their collections they provided the artefacts and implied judgements about what was significant; and they often stimulated the search for further acquisitions in the same area. Thus the Smithsonian's holdings in electric power were given greater validity in 1961 when it acquired the artefacts used by Malcolm MacLauren to illustrate his *The Rise of the Electrical Industry during the Nineteenth Century.* We came late to radio, and it would have been very difficult to have obtained a broad base without assistance from private collectors. The first of several was L. C. F. Horle in 1952, followed by a very large collection by Franklin Wingard a decade later. Most of the artefacts assembled by George Clark, lawyer and sometime historian for RCA, ended up at Dearborn, but his large and important archival collection came to the Smithsonian in 1959. Although we had acquired vacuum tubes from a variety of sources over the years, Paul Watson's collection gave a sense of comprehensiveness.

Figure 5. William J. Hammer in his study.

Among our more colorful benefactors was William J. Hammer. Associated with Edison at Menlo Park, he organized many of Edison's public exhibitions. He also was a compulsive collector. Especially impressive was his light bulb collection, which went to Henry Ford's museum at Dearborn. But the Smithsonian eventually (in 1962) acquired a significant number of items, many of them associated with Edison

There have also been the anonymous corporate collectors who, often with grudging (if any) support, managed to preserve significant examples of company history which eventually were presented to the Smithsonian (often in lieu of being dumped in the trash). Among the more important such collections are those from Weston Electrical Instrument Company in 1954, Western Union in 1972, and Texas Instruments in 1987. Each of these, significant in itself, stimulated further acquisitions in the area, respectively, of measuring instruments, telegraphy, and micro-electronics.

The last point can be made for even very small collections. In 1968 Priscilla Griffin de Moduit gave the museum an example of the first GE toaster (1908). At the time the Division of Electricity had virtually no electrical appliances. But this gift led directly to a campaign to assemble a modestly impressive number of such devices—ranging from hot plates

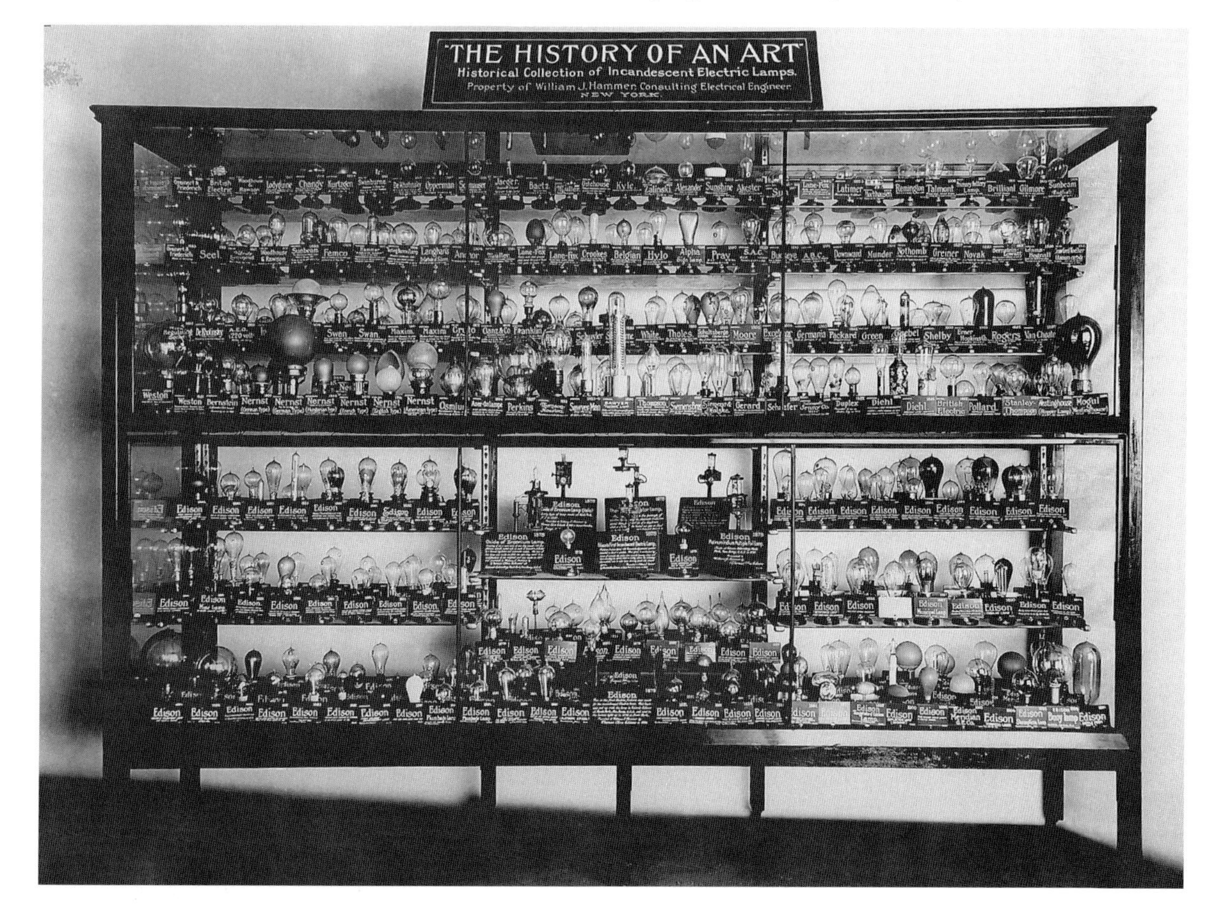

Figure 6. Hammer mounted his collection in several large cases. Most of the light bulbs, together with he cases, went to Henry Ford's museum in Dearborn. But a few found their way to the Smithsonian.

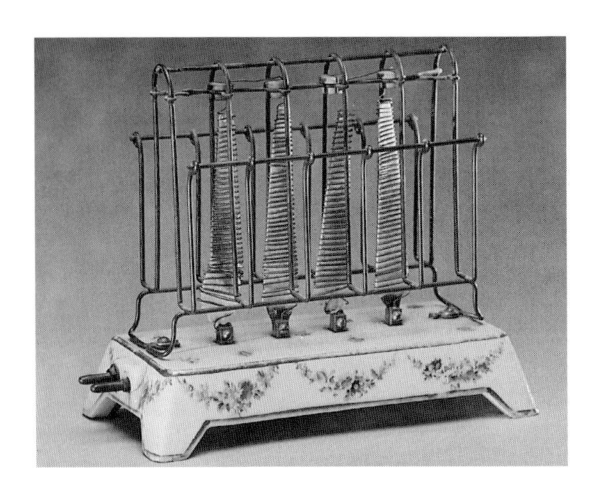

Figure 7. Mrs. de Moduit's 1908 GE toaster.

and coffee pots to washing machines and refrigerators. And this has proven to be of great assistance in helping the museum develop exhibits with broader social context.

What is true for Electricity is true for other collecting areas in the museum, including the Archives Center and the Library. And what is true here, by all indications, is true throughout the world. In their collections museums embrace the idiosyncracies, the character of their contributors. But in an era of increasing professionalism it can be difficult to see this in exhibits, which tend more and more to be political statements (in the academic as much as the social sense), and which use selected artefacts in the service of those statements. There are good reasons for continuing to display collections as collections. One is as a means of demonstrating the special origins of the particular museum. Another is to find in the passion of an individual collector evidence for a collective passion that a society can have for its past, especially as represented in artefacts.

* * *

Meters and motors, telegraphs and transistors, light bulbs and Leyden jars—there are a vast number of collectibles within the electrical ranks. In most museums they are grouped as electrostatics, power, telecommunications, microelectronics. Occasionally, as at the Smithsonian, these may all be combined in one happy curatorial family; but more frequently they are split off, either alone or with non-electric companions. Thus both the Science Museum and the Deutsches Museum have separate electric power sections, and they fold early electrostatics into physics. For the Science Museum, telecommunications is a stand-alone curatorial responsibility, while microelectronics and computers are linked; in the Deutsches Museum the three are connected together. The Musée des Arts et Métiers, which is joining us as the fourth sponsor of *Artefacts*, has been reconstituted with an expanded curatorial staff covering seven areas, with electrical collections divided among Scientific Instruments, Energy, and Communications. These differences clearly reflect differences in the institutional histories, including undoubtedly the influences of individual collectors.

With a few exceptions, other museums lack a staff of curators sufficiently large to allow them the luxury of separating subject areas beyond very large categories. This means that clues to the collecting history are probably not readily apparent from the administrative structure, though occasionally it is clear from the physical structure that some kind of merger has taken place (as, for instance, a new building that signals the joining of the Telemuseum to the Tekniska Museum in Stockholm in 1975). It would therefore require much more information than is readily available to pursue this argument in taking a broader view of the growth of technical museums.

Instead, let me close by noting that electricity is well represented in museums throughout the world. Following is a compoilaton of those that I have visited personally (some, admittedly, not for many years) or about which I have secondary knowledge. I hope that readers will be encouraged to visit some of them, as it is convenient, and to add personal nominations to the list.

They are separaated into groups. First is the large number (one might even say a surprisingly large number) of national museums of science and technology, all with significant electrical collections, often with scope extending beyond national borders. It should also be noted that not all of these encompass 'all' of science and technology. Indeed, telecommunications is a subject that is often left, at least for detailed examination, to another entity (more on this below). And transportation, because of the size of the objects, is also frequently treated elsewhere. The Smithsonian, for instance, has a separate museum for Air and Space. The Science Museum and Deutsches Museum have followed suit to the extent that they have established new structures for their aircraft collections, though retaining some portion in their home buildings. I might also note that the Smithsonian is unusual in incorporating science and technology into a general history museum. This junction was more or less clear in the original name, Museum of History and Technology. The current name, National Museum of American History, reflects an approach that attempts to integrate science and technology into broader historical concepts, though the collections themselves are organized in traditional fashion.

Second are museums that lack a national mandate but still have a collecting policy that is broad in subject matter but usually more restricted geographically. This focus can often make their exhibits more comprehensible, and more interesting, to the visitor.

Third is a small number of academically-oriented museums with narrowly conceived collections that are especially appealing to scholars.

Which is not to say that other museums don't also have academic connections or that these museums don't have exhibits that appeal to the general public.

Fourth are the telecommunications museums, which usually mark their origins from national postal-telegraph-telephone services though occasionally have been founded by other specialized interests.

Fifth are what I arbitrarily call specialized museums. Many of these are associated with individuals, others deal with narrow topics that are not conveniently included elsewhere.

1. National Museums.

Australia: Power House Museum (Sydney).
Austria: Technisches Museum (Vienna).
Canada: National Museum of Science and Technology (Ottawa).
China: China Museum of Science and Technology (Beijing).
Czech Republic: Narodni Technicke Muzeum (Prague).
Denmark: Teknisk Museum (Helsingor).
Finland: Tekniikan Museo (Helsinki).
France: Musée des Arts et Métiers, also known as the Musée National des Techniques, Conservatoire National des Arts et Métiers (Paris).
Germany: Deutsches Museum, or more fully as Deutsches Museum von Meisterwerken der Naturwissenschaft und Technik (Munich).
Great Britain. The Science Museum, more fully now known as the National Museum of Science and Industry (London).
Hungary: Hungarian Museum for Science and Technology (Budapest).
India: National Science Centre (Delhi).
Italy: Museo di Storia della Scienza e della Technica Leonardo da Vinci (Milan).
Japan: National Science Museum (Tokyo).
Mexico: Museo de Tecnologia (Mexico City).
Netherlands: Museum Boerhaave (Leiden).
Norway: Norsk Teknisk Museum (Oslo).
Poland: Muzeum Techniko NOT (Warsaw).
Romania: Muzeul Tehnic 'Prof. Ing. D. Leonida,' Naciolen Politechniceski Muzei (Sofia).
Russia: Polytechnic Museum (Moscow).
Scotland: National Museums of Scotland (Edinburgh).
Slovakia: Tecknicke Muzeum (Brno).
South Africa: Museum of Science (Johannesburg).
Sweden: Tekniska Museet (Stockholm).
Switzerland: Technorama da Schweiz (Winterthur).

2. Regional Museums:

Argentina: Museo Tecnologico 'Ingeniero Eduardo Latzina' (Buenos Aires).

Brazil: Museu de Ciencias (Sao Paulo).
Germany: Deutsches Technik Museum (formerly Museum für Verkehr und Technik) (Berlin) with collections related to energy and communications.
***: Landesmuseum für Technik und Arbeit (Mannheim).
Great Britain: The Museum of Science and Industry in Manchester has expanded considerably in recent years and has a broad collection of electrical apparatus, together with an extensive archive of the Electricity.
***: The Birmingham Museum of Science and Industry is located in quarters previously owned by Elkington & Co (electroplating) and features local technical history but has a significant eclectic collection of electrical items.
***: The Glasgow Art Gallery and Museum features technology of western Scotland, including a number of items related to William Thomson (Kelvin) (Glasgow).
***: The Science Museum in Newcastle includes important items from local pioneers Armstrong, Parsons and Swan.
India. Birla Industrial and Technological Museum (Calcutta). Although I have included the Delhi museum in the list of national technical museums, above, under the National Council of Science Museums responsibility for preserving the Indian technical heritage is shared with the Birla and Visvesvaraya museums.
***: Visvesvaraya Industrial and Technological Museum (Bangalore).
Japan: Chiba Museum of Science and Industry (Ichikawa-City) has a modest collection, including some electrical items.
United States: Henry Ford Museum and Greenfield Village (Dearborn) takes a national approach to its collecting, with a strong technical emphasis. Originally name the Edison Institute (and officially opening on the fiftieth anniversary of Edison's incandescent lamp) it holds a reconstruction of the Menlo Park compound and also of a Detroit 1880s generating station, together with much original apparatus. Other collections extend through the range of electrical technology.
***: New York State Museum (Albany), a general museum with strong technical emphasis.
***: South Carolina Museum (Columbia), a general museum with strong technical emphasis.

3. Academically-oriented museums.
England: Museum of the History of Science (Oxford), with a strong collection of electrostatic apparatus.
***: Museum of the Cavendish Laboratory (Cambridge), including apparatus associated with Maxwell and his successors.
Italy: Istituto e Museo di Storia della Scienza (Florence).

United States: Collection of Historical Scientific Instruments (Cambridge), including electrostatic instruments but also telephones and other items not necessarily linked to Harvard.
***: MIT Museum (Cambridge), with collections that are associated with MIT.

4. Telecommunications museums.
Argentina: Museo entel de Telefonia (Buenos Aires).
***: Museo Postal y Telegrafico 'Doctor Ramon J. Carcano' (Buenos Aires).
Austria: Post- und Telegraphen-Museum (Vienna).
Belgium: Musée des Postes et Télécommunications (Brussels).
Brazil: Museu de Telefone (Sao Paulo).
Canada: Bell Canada Telephone Historical Collection (Montreal).
***: Canadian Forces Communications and Electronics Museum (Kingston).
***: Telecommunications Museum of Canada (Brantford).
Chile: Museo Postal-Telegrafico (Santiago).
Denmark: Post- og Telegrafmuseet (Copenhagen).
***: Telefonmuseet (Copenhagen).
Finland: Posti-ja Telemuseo (Helsinki).
France: Centre National d'Etudes de Telecommunication (Issy).
***: Musée de Radio-France (Paris).
***: Musée d'Histoire des P.T.T. d'Alsace (Riquewihr).
***: Musée de la Poste et des Voyages (Amboise).
Germany: Deutsches Postmuseum (Frankfurt).
***: Deutsches Rundfunk-Museum (Berlin).
Great Britain: British Telecom Collection (Oxford).
***: British Telecom Museum (Taunton).
***: Museum of Communication (Edinburgh).
Hungary: Postamuzeum (Budapest).
Ireland: RTE Broadcasting Museum (Dublin).
Italy: Museo Storico PT (Rome).
Mexico: Muzeo Postal (Mexico City).
Netherlands: Het Nederlands Postmuseum (The Hague).
Switzerland: Schweizerisches PTT-Museum (Bern).
Tunisia: Musée National des PTT (Tunis).
United States: Antique Radio Relay League Museum of Amateur Radio (Newington CT).
***: Antique Wireless Association Electronic-Communication Museum (Bloomfield NY), a large collection of early wireless and radio apparatus.
***: Museum of Broadcast Communications (Chicago IL), mainly recordings and documentation with some apparatus).
***: Museum of Independent Telephony (Abilene KS), representing non-Bell equipment from the days before the breakup.

***: New England Museum of Wireless and Steam (East Greenwich, RI).
***: Pacific Bell Museum and Archives (San Francisco).
***: Pavek Museum of Broadcasting (Minneapolis).

5. Special museums

Austria: Schlossmuseum (Linz), somewhat uncomfortably in this category, but which fortuitously has a collection of electrical and physical artefacts from an 18th-century Jesuit college.
Canada: Alexander Graham Bell Museum (Baddeck), near Bell's summer home in Nova Scotia, where he did much experimentation, but lacks much in the way of electrical material.
***: Hearts Content Cable Station, at the Newfoundland terminus of the first successful Atlantic cable in 1866, equipped with early 20th-century apparatus.
***: Hydro Hall of Memory (Niagara Falls), in Ontario Hydro generating station.
***: Musée Historique d'Electricité Labadie (Longueuil).
France: Musée Ampère et de l'Electricité (Poleymieux), including a nice collection of early 19th-century apparatus.
***: Musée Branly (Paris), in his laboratory.
Germany: Deutsches Röntgen-Museum (Remscheid).
***: Elektromuseum Heinrich-Mayer-Haus (Esslingen).
***: Elektrotechnisches Museum der HASTRA (Hannover).
***: Electrum: das Museum der Elektrizität (Hamburg).
***: Siemens Forum (Munich), which has been converted in part from its earlier museum format but still with an impressive collection of artifacts and archival material.
Great Britain: The Royal Institution (London) maintains Faraday's laboratory and preserves and exhibits artefacts used by him and his successors.
Italy: Il Tempio Voltiano (Como), in town of Volta's birth and death, includes some 200 artefacts from the period, most attributed to Volta.
Japan: Tokyo Electric Power Company Electrical Museum (Tokyo), in formation, with a substantial collection of apparatus from TEPCO.
***: Yokogawa Museum of Measurement (Tokyo), in formation, with an international collection of electrical meters.
Netherlands: Teylers Museum (Haarlem), with large van Marum electrostatic machine and other turn-of-the-nineteenth-century apparatus.
United States: The Bakken Library and Museum (Minneapolis MN), which includes a broad range of items in its medical-electrical collections.
***: Bradbury Science Museum (Los Alamos NM), based on apparatus from Los Alamos Laboratory.
***: Computer Museum (Boston)

***: Discovery Museum, formerly Sacramento Museum of History, Science & Technology (Sacramento), included here because of a substantial collection of appliances and power-related items from the Madsen Electric Company.
***: Edison National Historic Site (West Orange, NJ), Edison's laboratory after 1886.
***: Edison Winter Home and Museum (Fort Myers FL), with emphasis on apparatus used in experiments at that location.
***: French Cable Station Museum in Orleans, terminus on Cape Cod of French Atlantic cables with early 20th-century equipment.
***: Georgetown Energy Museum (Georgetown CO), on site of generating plant.
***: Historical Electronics Museum (Baltimore, MD), with special emphasis on radar.
***: Motorola Museum of Electronics (Schaumburg IL).
***: Museum of Incandescent Lighting (Baltimore), a substantial collection of light bulbs.
***: National Cryptologic Museum (Fort Meade MD).
***: Schenectady Museum (Schenectady NY), local collections, including material from GE.
***: U. S. Army Signal Corps and Fort Gordon Museum (Augusta GA).
***: Western Museum of Mining and Industry (Fort Collins CO).
Yugoslavia: Tesla Museum (Belgrade) with a modest amount of apparatus, much in the way of personal effects, and a substantial archive.

Notes

1. For a psychologist's view, see Werner Meunsterberger, *Collecting, an Unruly Passion: Psychological Perspectives* (San Diego, New York, London, 1995)
2. autobiography 22–23.
3. For more examples, see John Elsner and Roger Cardinal, eds., *The Cultures of Collecting* (London: Reaktion Books, 1994)
4. For Darwin, see Duncan Park, 'The *Beagle* Collector and his Collections,' in David Kohn (ed.), *The Darwinian Heritage* (Princeton, 1985), pp. 973–1019; for Roosevelt, communication from Frank Bruns, National Postal Museum, Smithsonian Institution.
5. Columbia River Maritime Museum, 'Background Information,' July, 1996.
6. 'The Past and Future of the Textile Museum,' *Textile Museum Journal* 1 (Dec. 1963), p. 62
7. Descriptive booklet, *The Bakken: A Library and Museum of Electricity in Life* (St. Paul, 1986); the name has since been shortened to The Bakken Library and Museum.
8. Exhibit label, Baltimore Museum of Art, 1998.
9. Judith Sellers, 'Florence Dibell Bartlett: The Prime Mover,' in Richard Polese, ed., *Celebrate! The Story of the Museum of International Folk Art* (Santa Fe: The Museum of New Mexico Press, 1979), 10–13.
10. Paul Winkler, 'Girard Foundation Collection Sparks Expansion Plans,' ibid., 46–48.
11. Edward Miller, *That Noble Cabinet: A History of the British Museum* (Athens, Ohio: Ohio University Press, 1974); Carol Gibson-Wood, 'Classification and Value in a Seventeenth-Century Museum: William Courted's Collection,' *Journal of the History of Collections* 9, No. 1 (1997), 61–77; Arthur MacGregor, 'The Life, Character and Career of Sir Hans Sloane.' in Arthur MacGregor, ed., *Sir Hans Sloane: Collector, Scientist, Antiquary, Founding Father of the British Museum* (London: British Museum Press, 1994), 11–44; for the lottery see Marjorie Caygill, 'Sloane's Will and the Establishment of the British Museum,' ibid., 45–68.

12. 'A History of the Museums,' in Susan Stitt, et al, eds., *The Carriage Museum: Souvenir Publication* (Stony Brook: the Museums at Stony Brook, 1987), 19–37.
13. Walker, p. 30. *John Walker, National Gallery of Art* (New York: Harry N. Abrams, 1975; David Finley, *A Standard of Excellence: Andrew W. Mellon Founds the National Gallery of Art at Washington* (1973: Smithsonian Institution Press, 1973)
14. quoted in Annual Report, 1888, p. 3.
15. See Smithsonian Annual Reports for this period.

Index